Praise for *Life's Edge*

"Stories that both dazzle and edify . . . Particularly brilliant in telling the story of DNA . . . Zimmer is an astute, engaging writer—inserting the atmospheric anecdote where applicable, drawing out a scientific story and bringing laboratory experiments to life. This book is not just about life but about discovery itself. It is about error and hubris but also about wonder and the reach of science."

—Siddhartha Mukherjee, *New York Times Book Review*

"The pleasures of *Life's Edge* derive from its willingness to sit with the ambiguities it introduces, instead of pretending to conclusively transform the senseless into the sensible."

—*The Washington Post*

"A fascinating and well-written mapping of the edges of biology, which will have broad appeal to nonscientists."

—*Library Journal* (starred review)

"Diligently tackles the true definition of life . . . Zimmer invites us to observe, ponder, and celebrate life's exquisite diversity, nuances, and ultimate unity."

—*Booklist* (starred review)

"From the struggle to define when life begins and ends to the hunt for how life got started, [*Life's Edge*] offers an engaging, in-depth look at some of biology's toughest questions."

—*Science News*

"A master science writer explores the definition of life . . . An ingenious case that the answers to life's secrets are on the horizon."

—*Kirkus Reviews*

"Carl Zimmer shows what a great suspense novel science can be. *Life's Edge* is a timely exploration in an age when modern Dr. Frankensteins are hard at work, but Carl's artful, vivid, irresistible writing transcends the moment in these twisting chapters of intellectual revelation. Prepare to be enthralled."

—Jennifer Doudna, Nobel Laureate, coauthor of *A Crack in Creation*

"Profound, lyrical, and fascinating, *Life's Edge* will give you a newfound appreciation for life itself. It is the work of a master science writer at the height of his skills—a welcome gift at a time when life seems more precious than ever."

—Ed Yong, author of *I Contain Multitudes*

PREVIOUSLY GARNERED PRAISE FOR CARL ZIMMER

"One of the best science writers we have today."

—Rebecca Skloot, author of *The Immortal Life of Henrietta Lacks*

"No one unravels the mysteries of science as brilliantly and compellingly." —David Grann, author of *Killers of the Flower Moon*

"Nobody writes about science better."

—Neil Shubin, author of *Your Inner Fish*

"Carl Zimmer makes the complex science of heredity read like a novel." —Elizabeth Kolbert, author of *The Sixth Extinction*

ALSO BY CARL ZIMMER

She Has Her Mother's Laugh

A Planet of Viruses

Evolution: Making Sense of Life

The Tangled Bank

Science Ink

Brain Cuttings

The Descent of Man: The Concise Edition

Microcosm

Soul Made Flesh

Evolution: The Triumph of an Idea

Parasite Rex

At the Water's Edge

LIFE'S EDGE

The Search for What It Means to Be Alive

CARL ZIMMER

DUTTON

DUTTON

An imprint of Penguin Random House LLC
penguinrandomhouse.com

Previously published as a Dutton hardcover in March 2021
First Dutton trade paperback printing: March 2022

LIBRARY OF CONGRESS CATALOGING-IN-PUBLICATION DATA

Names: Zimmer, Carl, 1966– author.
Title: Life's edge : the search for what it means to be alive / Carl Zimmer.
Description: New York : Dutton, an imprint of Penguin Random House LLC, [2021] |
Includes bibliographical references and index.
Identifiers: LCCN 2020039762 (print) | LCCN 2020039763 (ebook) |
ISBN 9780593182710 (hardcover) | ISBN 9780593182727 (ebook)
Subjects: LCSH: Life sciences. | Life (Biology)
Classification: LCC QH311 .Z56 2021 (print) | LCC QH311 (ebook) | DDC 570—dc23
LC record available at https://lccn.loc.gov/2020039762
LC ebook record available at https://lccn.loc.gov/2020039763
Dutton trade paperback ISBN: 9780593182734

Printed in the United States of America
1st Printing

BOOK DESIGN BY LORIE PAGNOZZI

To Grace, my love and life

CONTENTS

PART THREE

PART FOUR

INTRODUCTION

THE BORDERLAND

In the fall of 1904, the Cavendish Laboratory was full of curious experiments. Clouds of mercury shuddered with flashes of blue light. Lead cylinders pirouetted on copper disks. The ivy-covered building on Free School Lane, nestled in the heart of Cambridge, was the most exciting place for physicists to be, not just in England but in the entire world, a place where they could toy with the fundamental pieces of the universe. Amidst this forest of magnets and vacuums and batteries, it would have been easy to overlook one small experiment sitting forlornly by itself. It consisted of little more than a glass tube capped with cotton, half-filled with a few spoonfuls of brown broth.

But something was coming into being in that tube. In a few months the world would collectively gasp at it. Newspapers would celebrate the experiment as one of the most remarkable achievements in the history of science. One reporter would describe what lurked in the tube as "the most primitive form of life—the 'missing link' between the inorganic and organic worlds."

This most primitive life was the creation of a thirty-one-year-old physicist named John Butler Burke. In photographs from around the time of the experiment, Burke's boyish face has a melancholy cast. He was born in Manila to a Filipino mother and an Irish father. As a boy he traveled to Dublin for schooling and ended up at Trinity College, where he studied X-rays, dynamos, and the mysterious sparks released by sugar. Trinity awarded Burke a gold medal in physics and chemistry. One professor described him as "a man who is gifted with the power of exciting in others the enthusiasm which he brings to bear upon his own lines of investigation." After finishing his studies, Burke moved from Dublin to England to teach at a series of universities. His father soon died and his mother—"an old lady of very large means," as Burke later called her—supported him with a generous allowance. In 1898, Burke joined the Cavendish.

Nowhere on Earth had physicists learned so much in so little time about matter and energy. Their most recent triumph, courtesy of the lab's director, Joseph John Thomson, was the discovery of the electron. In his first few years at the Cavendish, Burke followed up on Thomson's work by running experiments of his own on the mysterious charged particles, investigating how electrons could light up clouds of gas. But then a new mystery lured him away. Like many other young physicists at the Cavendish, Burke started experimenting with a glowing new element called radium.

A few years beforehand, in 1896, a French physicist named Henri Becquerel had discovered the first evidence that ordinary matter could cast off a strange form of energy. When he wrapped uranium salts in a black cloth, they created a ghostly image on a photographic plate nearby. It soon became clear that the uranium was steadily releasing some kind of potent particle. To follow up on Becquerel's work, Marie and Pierre Curie extracted uranium from an ore called pitchblende. In the process, they discovered that some of the energy was coming from a second element. They named it radium and christened its new form of energy "radioactivity."

Radium unleashed so much energy that it could keep itself warm. If scientists set a piece atop a block of ice, it could melt its own weight in water. When the Curies mixed radium with phosphorus, the particles unleashed by the radium made the phosphorus glow in the dark. As news of this rare, exotic substance spread, it became a sensation. In New York, dancers put on radium-coated outfits to perform in darkened casinos. People wondered if radium would become a mainstay of civilization. "Are we about to realize the chimerical dream of the alchemists—lamps giving light perpetually without the consumption of oil?" one chemist mused. Radium also seemed to have a vitalizing power. Gardeners sprinkled it on their flowers, convinced it could make them grow bigger. Some people drank "liquid sunshine" to cure all manner of ills, including even cancer.

It was cancer that would eventually claim Marie Curie's life in 1934, probably because of the radium and other radioactive elements she worked with on a daily basis. Now that we understand the deadly risk posed by radioactivity, it's hard to imagine how anyone could think that radium could have vitalizing powers. But in the early 1900s scientists knew surprisingly little about the nature of life. The best they could say was that its essence lurked in the jellylike substance in cells, a material they called protoplasm. It somehow organized cells into living things and was passed down from one generation to the next. Beyond that, little was certain and all manner of ideas were viable.

To Burke, life and radioactivity displayed a profound similarity. Like a caterpillar becoming a moth, a radium atom could undergo a transformation that seemed to come from within. "It changes its substance—in a limited sense it lives—and yet it is ever the same," Burke declared in a 1903 magazine article. "The distinction, apparently insuperable, that the biologist holds to exist between living and so-called dead matter, should thus pass away as a false distinction . . . All matter is alive—that is my thesis."

Burke said all this as a scientist, not a mystic. "We must be careful lest our imagination should carry us away, and lead us into regions of pure fancy, to a height beyond the support of experimental facts," he warned. To prove his thesis, Burke designed an experiment: he would use radium to create life from lifeless matter.

To carry out this act of creation, Burke prepared some bouillon, cooking chunks of beef in water and sprinkling in salt and gelatine. Once the ingredients had turned to a broth, he poured some into a test tube and put it over a flame. The heat destroyed any cow cells or microbes that might be lurking in the liquid. All that was left was a sterile bouillon made up of loose, lifeless molecules.

Burke put a pinch of radium salt in a tiny sealed vial, which was suspended over the broth. A platinum wire wrapped around the vial and snaked out a side port. To launch the experiment, Burke pulled the free end of the wire until the vial cracked. The radium tumbled into the broth below.

After he let the radioactive broth stew overnight, Burke saw that it had changed: a cloudy layer had formed on its surface. Burke drew off a little to see if it was made of contaminating bacteria. He spread it over a petri dish loaded with food for microbes. If the cloudy layer had any bacteria in it, they would feast until they grew into visible colonies.

But no colonies formed. Burke concluded the layer must have been formed by something else. Taking another sample of the cloudy layer, he spread it on a glass slide and put it under a microscope. Now he could see that it contained a scattering of specks far smaller than bacteria. A few hours later, when he checked again, the specks had vanished. But the next day they returned, and Burke began drawing them, documenting how they grew in size and changed in shape. Over the course of the next few days they turned into spheres, with inner cores and outer rinds. They stretched into dumbbells. They bulged and pinched into miniature flowers. They divided. And then, after two weeks, they fell apart. Some might say they died.

As Burke sketched these changing shapes, he could tell they were not bacteria. It wasn't just that they were too small. When Burke put some of them in water, they dissolved away—a fate that bacteria did not suffer. Yet Burke was convinced these radium-laced blobs were not crystals or any other familiar forms of lifeless matter. "They are entitled to be classed among living things," Burke concluded. He had created "artificial life," as he called it—creatures that existed at the far edges of life's territory. And to these things he gave a name that commemorated the element that gave birth to them: radiobes.

Burke could only guess at how he had created his radiobes. When he dropped the radium into the broth, the element must have given the molecules the powers of growth, organization, and reproduction. "The constituents of protoplasm are in the bouillon," he later wrote, "but the vital flux is in the radium."

That December the scientists of the Cavendish Laboratory celebrated Burke's discovery at their annual dinner in a back room at a Cambridge restaurant. Dressed in black tie, they read lyrics written by a physicist named Frank Horton. They belted out "The Radium Atom" to the tune of an old music hall song:

Oh, I am a radium atom,
In pitchblende I first saw the day,

But soon I shall turn into helium:
My energy's wasting away.

The physicists sang about the gamma rays and beta rays that radium unleashed, and then they turned to Burke's experiment:

Through me they say life was created
And animals formed out of clay,
With bouillon I'm told I was mated
And started the life of today.

Five months later, on May 25, 1905, Burke published his first report on radiobes in the journal *Nature*. He adorned his account of his experiment with three blurry sketches of "highly organized bodies." Burke ended his report by christening the bodies as radiobes, thus "indicating their resemblance to microbes, as well as their distinct nature and origin," he said.

When the reporters came calling, Burke at first shied away from claiming too much for his discovery. But they gnawed at his resolve like termites in old wood. Pointing out that radioactive minerals were turning out to be surprisingly widespread, Burke speculated that radiobes existed across the entire planet. "Life may have originated on earth in that way," he told one reporter.

The public lapped it up. "Has Radium Revealed the Secret of Life?" the *New York Times* asked. Burke's radiobes, they marveled, seem to "tremble between the inertia of inanimate existence and the strange throb of incipient vitality."

The news made Burke as famous as his radiobes. "John Butler Burke has suddenly become the most talked about man of science in the United Kingdom," the *New York Tribune* reported. The *Times* of London anointed him "one of the most brilliant of our younger physicists," who had carried out "one of the supremely great achievements of all time." Another British writer judged that "Mr. Burke attained suddenly to a notoriety which, in this country, is usually reserved for prominent athletes." Letters full of questions about the radiobes arrived "from the remotest corners of the Earth," Burke later recalled.

Burke enjoyed his fame. Instead of running more experiments at the Cavendish, he traveled from lecture hall to lecture hall showing off his lantern slides. Magazines paid him handsomely for his words. The *World's Work* went so far as to compare Burke to Darwin. Radiobes "provoked more discussion, perhaps, than any event in the history of science since the publication of the *Origin of Species*," they declared. In 1859, Charles Darwin had laid out a theory of how life evolved. Now, almost a half century later, Burke was wrestling with an even greater mystery: life itself. Chapman and Hall, one of London's leading publishers, gave Burke a contract to write a book about his theory. *The Origin of Life: Its Physical Basis and Definition* came out in 1906.

Whatever caution Burke originally had was now gone. In his book, he held forth on the properties of living matter, on the "borderland between mineral and vegetable kingdoms," on enzymes and nuclei, on his own electric theory of matter, and on something he called "mind-stuff." Burke unhelpfully described mind-stuff as "perception in the universal mind which constitutes the 'great ocean of thought' in which we live and move and have our being."

And with those words Burke reached his Icarus peak. Soon a wave of brutal reviews of *The Origin of Life* came out, scoffing at Burke's hubris. Here was a physicist holding forth on the nature of life when he didn't even know the difference between chlorophyll and chromatin. "Biology is decidedly not his forte," one reviewer sniffed.

An even more devastating verdict soon came from a fellow scientist. W. A. Douglas Rudge, who had also worked at the Cavendish for a few years, decided to run Burke's radiobe experiments for himself. He recognized ways to make them more rigorous—running separate trials with tap water and distilled water, for example. Instead of Burke's "mere drawing," as Rudge called it, he documented his results with photographs. When Rudge cooked his broth with distilled water, he discovered, the radium produced nothing. In tap water, Rudge found some odd shapes, but no sign of the lifelike radiobes Burke had drawn.

Burke tried to smear Rudge as an amateur, but other scientists saw his report to the Royal Society as the final word on radiobes. "Mr. Rudge has carried out the experiments that Mr. Burke should have made long ago," declared Norman Robert Campbell, a physicist at the Cavendish. "Mr. Rudge has produced convincing evidence that the 'cells,' or radiobes, are nothing but little bubbles of water produced in the gelatin by the action of the salts upon it."

In September 1906, Campbell published a vicious attack on Burke. It was ostensibly a review of *The Origin of Life*, but it read more like a character assassination. "Mr. Burke was not educated at Cambridge; he had been at two universities before he came thither as an advanced student," Campbell scoffed. "It is misleading to say, in connection with his recent publications, that Mr. Burke is 'of the Cavendish Laboratory.' He did some physical research there a few years ago: during his investigations of the biological properties of his radiobes he merely stored in the room in which he had done his former work some of the test-tubes in which those bodies were 'incubating.'"

It was around this time that Burke stopped working at the Cavendish. Whether he quit or was barred, no one can say. In December 1906 the lab gathered again for another end-of-the-year dinner. They had cause to celebrate: Thomson had just won the Nobel Prize. But the song for 1906 was not an ode to the electron. Instead, the mathematician Alfred Arthur Robb wrote a song set to the tune of "The Amorous Goldfish" from the 1896 musical *The Geisha*.

It was entitled "The Radiobe."

A radiobe swam in a bowl of soup
As dear little radiobes do,
And Butler Burke gave a wild war whoop
As he over his microscope did stoop,
And it came in the field of view.
He said: "This radiobe clearly shows
How all the forms of life arose;
And further plainly shows," said he,
"What a very great man is J.B.B.!"

In the years that followed, Burke took a long fall—one that only ended with his death forty years later in 1946. After he left the Cavendish, no one offered him a plum professorship. Magazines lost interest in his ideas. He wrote two sprawling manuscripts but struggled for years to find a publisher. His income from lectures and writing dried up at the same time that his mother slashed his allowance. During World War I, Burke managed to support himself with a job inspecting airplanes, but after a few months poor health forced him to quit. In 1916 he begged the Royal Literary Fund for a loan to save him from "the dreaded event of bankruptcy." They turned him down.

As a young man, Burke had seemed on the verge of defining life, of charting its borders. But life got the better of him. In 1931, a quarter century after his brief fame, he published a dubious magnum opus, *The Emergence of Life*. It was a rambling mess. "Burke had gone right off the deep end," the historian Luis Campos later wrote. In the book, Burke flirted with levitation and other psychic phenomena. He remained fiercely loyal to his radiobes, which the world had long forgotten. He argued that life emerged from what he called "time-waves" that flowed between units of mind that make up the universe.

The more Burke thought about life, the less he understood it. At one point in *The Emergence of Life*, he offered a definition of life, but it sounded more like a cry for help: "Life is what IS."

I never learned about Burke when I was growing up. I was taught the standard pantheon of biologists, which is mostly made up of scientists with ideas that turned out to be right: Darwin and his tree of life, Mendel and his genetic peas, Louis Pasteur and his disease-causing germs. It's easier that way: to leapfrog from one designated hero to the next—to ignore the mirages along the way, the failures, the fame that curdled.

When I started writing about biology, I still didn't learn about Burke. I have had the good fortune to get to know many forms of life and many of the scientists who study them. I've hauled hagfish out of the North Atlantic, hiked into North Carolina longleaf pine forests to find Venus flytraps in the wild, and spotted orangutans lounging high in the canopies of Sumatran jungles. Scientists have shared with me what they've learned about the marvelous slime that hagfish make, the insect-destroying enzymes in carnivorous plants, the tools orangutans fashion out of sticks.

The beams of their scientific flashlights are bright, but only because they are narrow. Someone who spends her life tracking orangutans doesn't have enough time to become an expert on Venus flytraps. Venus flytraps and orangutans have something profoundly important in common—they are alive—and yet asking biologists about what it means for something to be alive makes for an awkward conversation. They'll demur, stammer, or offer a flimsy notion that crumbles under even a little scrutiny. It's just not something that most biologists give much thought to in their day-to-day work.

This reluctance has long mystified me, because the question of what it means to be alive has flowed through four centuries of scientific history like an underground river. When natural philosophers began contemplating a world made of matter in motion, they asked what set life apart from the rest of the universe. The question led scientists to many discoveries but also many blunders. Burke was hardly alone. For a brief time in the 1870s, for example, many scientists came to believe that the entire ocean floor was carpeted with a layer of throbbing protoplasm. More than 150 years later, despite all that biologists have learned about living things, they still cannot agree on the definition of life.

Puzzled, I set out on a trip. I started out in the heart of life's territory: in the confidence that each of us has that we are alive, that we have a life that is bounded by birth on one side and by death on the other. Yet we feel our own life more strongly than we understand it. We know that other things are alive, too, like snakes and trees, even if we can't ask them. Instead, we rely on the hallmarks that all living things seem to have. I took a tour of these hallmarks, getting to know creatures that display them in their most impressive, most extreme forms. Eventually my travels took me out to life's edge, to the foggy borderland between the living and the nonliving, where I encountered peculiar things with some of life's hallmarks but not others. It was here, at last, that I first encountered John Butler Burke and came to appreciate that he deserves a place in our memory. It was here that I met his scientific descendants who still grope their way around life's edge, trying to figure out how life began or how weird it might get on other worlds.

Someday humanity may draw a map that will make this journey easier. In a few centuries, people may look back at our understanding of life and wonder how we could have been so blinkered. Life today is like the night sky four centuries ago. People gazed up at mysterious lights that wandered, streaked, and flared across the dark. Some astronomers at the time were getting the first inklings of why the lights traced their particular paths, but many of the explanations of the day would turn out to be wrong. Later generations would look up and instead see planets, comets, and red giant stars, all governed by the same laws of physics, all manifestations of the same underlying theory. We don't know when a theory of life might arrive, but we can hope, at least, that our own lives last long enough to let us see it.

PART ONE THE QUICKENING

THE WAY THE SPIRIT COMES TO THE BONES

As I made my way down the hairpin road, a sage brush–studded wall of sand to my right, I felt keenly aware of my own life. I could feel the steep slope in my legs. After a series of tight turns, the wall swung away, revealing a long, desolate beach. It ran northward, a sash of coast between high, slumping cliffs and the Pacific. Out over the sea, the sun hid behind clouds, a sky-wide bank of white. Earlier that day, in my hotel room, my phone had informed me the sky was cloudy and the temperature was in the low seventies. My brain responded to that information by choosing a light, long-sleeved shirt for my walk to the beach. And now my brain was updating its decision without cc'ing my conscious self.

Nerves sprinkled throughout my skin sensed the humidity and temperature of the layer of air encasing my body. Voltage spikes traveled from the nerve endings along long branches known as dendrites until they reached the cores of the nerves, called the somas. From there, new signals raced onward along long, cable-shaped extensions called axons. The axons reached my spine and traveled up toward my head. From neuron to neuron, the signals from the outside world made their way into my brain and finally to a nub of neurons deep inside my skull.

Those neurons combined the Morse code readout from across my body to generate new, different signals. They carried commands instead of sensations. The new voltage spikes left my brain along outward-bound axons, through my brain stem and down my spinal cord, until they reached millions of glands in my skin. There, they created electric charges in twisted tubes that wrung water out of the surrounding cells. Sweat ran down my back.

My conscious self was annoyed with the brain that generated it. One of the few shirts I had brought with me was now drenched in salt water. I could not actually sense the trill of voltage spikes that shuttled information from skin to brain. I didn't feel a surge of blood in the center of my head as the heat-regulating part of my brain swung into action. In the moment, by the sea, I simply felt myself sweating. I felt annoyed. I felt alive.

As I felt aware of my own life, I also recognized other lives on the beach. A man walked lazily south, carrying a white-and-blue surfboard. Far to the north, a paraglider launched off from the top of the cliffs. The corkscrewing of the yellow paraglider wing spoke of intentions that arose in some human's brain and produced signals to hands gripping brake handles.

Along with human life, I could see feathered life as well. Sandpipers skittered along the surf. Their seed-sized brains sensed the flash of incoming waves and the cold foam around their legs, contracting muscles to keep their bodies upright, to scuttle to higher ground, to poke the sand for buried snails. The snails didn't quite have brains but rather fretworks of nerves that produced signals of their own for slowly, relentlessly burying their bodies into the earth. I contemplated the thousands of other subterranean nervous systems inside the mud dragons and the Pismo clams and other creatures buried below my feet. Out in the ocean, down the underwater canyon, other brains were swimming, carried along inside the buoyant bodies of leopard sharks and stingrays while the nerve nets of jellyfish drifted by.

After a few minutes of walking along the water, I stopped and looked down. A gigantic neuron, six feet long, lay on the sand. Most of it was made up of a glistening, caramel-colored axon. It curved gently like a heavily insulated electric cable. At one end it swelled into a bulb-shaped soma, which was crowned in turn by branches of dendrites. It could have been all that survived from a kraken that died in a battle with a pod of killer whales somewhere between here and Hawaii.

This fantastical neuron was, in fact, a stalk of elk kelp. It had washed up from an underwater forest a mile out to sea. What I had imagined to be an axon was the kelp's stipe, a trunk that not long ago anchored the organism to the ocean floor. What looked like a neuron's soma was in reality a gas-bloated bladder that kept the kelp upright in the ocean currents. The branching dendrites were the elk kelp's antlers, on which long blades had once grown. And the blades acted like the leaves of plants, catching what little sunlight filtered down through the seawater and fueling the growth of the elk kelp to heights that rivaled the palm trees that crowned the cliffs behind me.

The kelp had the kind of complexity that marks living things. But as I looked down at it, I could not say whether this particular kelp was still alive. I couldn't ask it how its day was going. It had no heartbeat I could check, no lungs to lift and lower a chest. But the kelp still glistened, its surfaces intact. Even if it could no longer capture sunlight, its cells might still be carrying on, using up its remaining fuel to repair its genes and membranes. At some point, maybe today or next week, its death would become certain.

But along the way, it would also become a part of life on land. Microbes would feast on its tough cuticles. Beach hoppers and kelp flies would follow, nibbling on its tender tissues. These wrack-feasting creatures would themselves become food for the sandpipers and terns. Plants would be fertilized by the kelp's nitrogen soaking into the ground. And a sweaty human being, his brain packed with thoughts of brains on this beach, would carry away in his neurons a memory of the kelp's neuron-like body.

The next morning I walked along the tops of the cliffs. North Torrey Pines Road cut north through La Jolla, California, alongside groves of looming tower cranes. With a stream of rush-hour traffic flowing by me, it was hard to remember the ribbon of wild coast tucked away close by. I crossed a eucalyptus-lined parking lot to get to the Sanford Consortium for Regenerative Medicine, a complex of glassed-in labs and offices. Once inside, I found my way to a third-floor laboratory, and there I met a scientist named Cleber Trujillo—Brazilian-born, with a close-cropped beard. Together we suited up in blue gloves and smocks.

Trujillo led me to a windowless room banked with refrigerators, incubators, and microscopes. He extended his blue hands to either side and nearly touched the walls. "This is where we spend half our day," he said.

In that room Trujillo and a team of graduate students raised a special kind of life. He opened an incubator and picked out a clear plastic box. Raising it above his head, he had me look up at it through its base. Inside the box were six circular wells, each the width of a cookie and filled with what looked like watered-down grape juice. In each well a hundred pale globes floated, each the size of a housefly head.

Every globe was made up of hundreds of thousands of human neurons. Each had developed from a single progenitor cell. Now these globes did many of the things that our own brains do. They took up nutrients in the grape-juice-colored medium to generate fuel. They kept their molecules in good repair. They fired electrical signals in wavelike unison, keeping in sync by exchanging neurotransmitters. Each of the globes—which scientists call organoids—was a distinct living thing, its cells woven together into a collective.

"They like to stay close to each other," Trujillo said as he looked at the undersides of the wells. He sounded fond of his creations.

The lab where Trujillo worked was led by another scientist from Brazil named Alysson Muotri. After Muotri immigrated to the United States and became a professor at the University of California at San Diego, he learned how to grow neurons. He took bits of skin from people and gave them chemicals that transformed them into embryo-like cells. Dousing them with another set of chemicals, he steered them to develop into full-blown neurons. They could form flat sheets covering the bottom of petri dishes, where they could crackle with voltage spikes and trade neurotransmitters.

Muotri realized that he could use these neurons to study brain disorders that arose from mutations. Instead of carving out a piece of gray matter from people's heads, he could take skin samples and reprogram them into neurons. For his first study, he grew neurons from people with a hereditary form of autism called Rett syndrome. Its symptoms include intellectual disability and the loss of motor control. Muotri's neurons spread their kelp-like branches across petri dishes and made contact with each other. He compared them to the neurons he grew from skin samples taken from people without Rett syndrome. Some differences leaped out. Most noticeably, the Rett neurons grew fewer connections. It's possible that the key to Rett syndrome is a sparse neural network, which changes the way signals travel around the brain.

But Muotri knew very well that a flat sheet of neurons is a far cry from a brain. The three pounds of thinking matter in our heads are a kind of living cathedral, if a cathedral were built by its own stones. Brains arise from a few progenitor cells that crawl into what will become an embryo's head. They gather together to form a pocket-shaped mass and then multiply. As the mass grows, it extends long, cable-like growths out in all directions, toward the forming walls of the skull. Other cells emerge from the progenitor mass and climb up these cables. Different cells stop at different points along the way and begin growing outward. They become organized into a stack of layers, known as the cerebral cortex.

This outer rind of the human brain is where we carry out much of the thinking that makes us uniquely human—where we make sense of words, read inner lives on people's faces, draw on the past, and plan for the distant future. All the cells that we use for these thoughts arise in a particular three-dimensional space in our heads, awash in a complex sea of signals.

Fortunately for Muotri, scientists came up with new recipes to coax reprogrammed cells to multiply into miniature organs. They made lung organoids, liver organoids, heart organoids, and—in 2013—brain organoids. Researchers coaxed reprogrammed cells to become the progenitor cells for brains. Provided with the right signals, those cells then multiplied into thousands of neurons. Muotri recognized that brain organoids would profoundly change his research. A disease like Rett syndrome starts reworking the cerebral cortex from the earliest stages in the brain's development. For scientists like Muotri, those changes happened inside a black box. Now he could grow brain organoids in plain view.

Together, Muotri and Trujillo followed the recipes that other scientists laid down for making organoids. Then they began creating recipes of their own to make a cerebral cortex. It was a struggle to find the blend of chemicals that could coax the brain cells onto the right developmental path. The cells often died along the way, tearing open and spilling out their molecular guts. Eventually the scientists found the correct balance. They discovered to their surprise that once the cells set off in the right direction, they took over their own development.

No longer did the researchers have to patiently coax the organoids to grow. The clumps of cells spontaneously pulled away from each other to form a hollow tube. They sprouted cables that branched out from the tube, and other cells traveled along the cables to form layers. The organoids even grew folds on their outer surface, an echo of our own wrinkled brains. Muotri and Trujillo could now make cortex organoids that would grow to hundreds of thousands of cells. Their creations stayed alive for weeks, then months, then years.

"The most incredible thing is that they build themselves," Muotri told me.

On the day I visited Muotri's lab, he was checking in on some organoids he had sent into space. He sat in his office, a glass box perched out on a balcony next to the lab. Muotri had a gentle, relaxed manner, as if he might at any moment take off early from work, scoop up the scarred surfboard leaning against the wall by his desk, and head for the water. But today he was focused on the most extravagant of his many experiments. Outside his window, the paragliders were taking flight in the distance. He paid them no mind. Aboard the International Space Station, 250 miles above Muotri's head, hundreds of his brain organoids were sitting inside a metal box. He wanted to know how they were faring.

For years astronauts aboard the space station had run experiments to see how cells grow in low Earth orbit. As they free-fell around the planet, the cells no longer experienced the same tug of gravity that has pulled on all life on Earth for the past 4 billion years. Strange things happen in microgravity, it turned out. In some experiments, the cells grew faster than they would on the ground. They sometimes became bigger. Muotri was curious to see if his organoids would grow into larger clusters in space and perhaps become more like our own brains.

When they won approval from NASA, Muotri, Trujillo, and their colleagues began collaborating with engineers to build a home for organoids in space. They designed an incubator that could nurture the organoids, keeping the conditions right for their development. A few weeks before I visited the lab, Muotri had poured a fresh batch of miniature brains into a vial, which he put in a backpack. Standing in the security line at San Diego International Airport, he had no idea what he'd say if anyone asked what was in the tube. *These are a thousand miniature brains I've grown in my lab, and I'm about to send them into space.*

Apparently, organoids don't grab that kind of attention. Muotri managed to board his flight without getting questioned. When he got to Florida, he handed the tube over to the engineers for a flight aboard a supply rocket. A few days later Muotri watched the SpaceX Falcon 9 rise from the earth.

When the payload arrived at the space station, the astronauts grabbed the box loaded with organoids and plugged it into a bay. There it sat for a month. When the experiment was over, the astronauts would dunk the organoids in alcohol. They would die, but their lives would be frozen at the moment of death. Once they fell back into the Pacific, were fished out, and were delivered to Muotri's lab, he'd be able to inspect their cells and see which genes they had used in space.

The entire effort depended on the organoids surviving till their appointed end, and Muotri didn't know whether they'd make it. To keep track of the organoids during their month in space, he had arranged for miniature cameras to spy on them, taking pictures every thirty minutes. The space station transmitted the photos down to Earth, and eventually Muotri could log into a remote server to grab them.

When he downloaded the first batch of images from early in the mission, they turned out to be a mess. Air bubbles blocked his view. For three weeks he had no idea how his organoids were doing. Now I watched Muotri connect to the server once more. He found a new image from the space station to download. The massive file decompressed and the picture appeared, stripe by stripe, on his screen.

"Oh!" Muotri called out. He laughed in disbelief. "I can actually see them!"

He moved his face close to the screen to inspect the image. Half a dozen gray spheres floated on a beige background.

"Yeah, they all look quite good," he said. "They're rounded, and they more or less have the same size. You don't see them fusing or clustering together." He rolled his chair back from his computer. "So this is all good news. I'm happy. This is fantastic."

Even in space, Muotri could tell his organoids were alive.

In late 2015, Muotri and Trujillo got their first chance to use their organoids to learn something about brains. In Brazil, doctors were struggling to understand why the brains of thousands of babies were drastically deformed. Their cerebral cortices were practically missing. It turned out that their mothers had been infected by a mosquito-borne virus called Zika, which had never before been found in the Americas. Muotri and Trujillo got a supply of Zika viruses and began infecting brain organoids. They wondered if they'd see a change.

"It was night and day," Muotri told me.

The Zika viruses immediately destroyed progenitor cells in young organoids. Without those cells, an organoid could not sprout cables to build a cortex. The experiments revealed that Zika viruses do not kill the cerebral cortex, so much as prevent it from growing in the first place. Once the scientists figured out how Zika viruses wreak their havoc, they were able to discover drugs that could block them. Those drugs then went into tests on animals to see if they could help prevent brain damage.

Word spread that Muotri was growing brain mimics by the thousands. Graduate students and postdoctoral researchers wanted in. When they joined his lab, they first had to train with Trujillo for months, learning the fine art of making organoids. I asked a graduate student named Cedric Snethlage to describe his education. Making a brain organoid wasn't just a matter of reading temperatures and pH levels off a protocol, he explained. Snethlage had to learn how to carry out each step by intuition—how far, for example, to tilt the wells to keep the organoids from sticking to the bottom. I told Snethlage that he sounded like he had just gone through cooking school.

"It's more like making a soufflé than making chili," he said.

Snethlage wanted to learn how to grow organoids to study neurological disorders. Other graduate students had come to Muotri's lab to discover how to make organoids more brain-like. Brain cells need nutrients and lots of oxygen to thrive, and the ones at the center of an organoid can starve. So some of Muotri's students were adding new cells to organoids that could develop into artery-like tubes. Others were adding immune cells to see if they might sculpt the branches of the neurons into more natural shapes.

Meanwhile, Cleber Trujillo's wife, Priscilla Negraes, began listening to the chatter going on between the organoid cells.

When a brain organoid reaches a few weeks in age, its neurons become mature enough to generate spikes of voltage. Those spikes can travel down an axon and trigger neighboring neurons to fire as well. Negraes and her colleagues created an eavesdropping device that could pick up the crackle. At the bottom of miniature wells, they placed eight-by-eight grids of electrodes. They filled the wells with broth and rested an organoid atop each array.

On her computer, the readout from the electrodes formed a grid of sixty-four circles. Whenever one of the electrodes detected a firing neuron, its circle swelled, turning from yellow to red. Week after week the circles reddened and swelled more often, but there was no pattern Negraes could see to the bursts. The cells in the organoids spontaneously fired on their own from time to time, creating neurological static.

But as the organoids got more mature, Negraes thought she saw some order emerge. Sometimes a few of the circles would all suddenly swell red together. Eventually all sixty-four electrodes registered signals at once. And then Negraes began to see them turn on and off in what looked like waves.

Was Negraes seeing actual brain waves developing in the organoids? She wished that she could compare the patterns she was seeing in her wells with the developing brains of human fetuses. But scientists had yet to figure out how to detect their electrical activity in the womb. The closest that anyone had managed was to study babies born premature, putting miniature EEG caps on their orange-sized heads.

Negraes and her colleagues enlisted a University of California, San Diego, neuroscientist named Bradley Voytek and his graduate student Richard Gao to compare organoids to premature babies. The earliest-born babies, with the least developed brains, produced sparse bursts of brain waves separated by long spells of disorganized firing. The babies that were born closer to term had shorter lulls, their bursts of brain waves growing longer and more organized. Organoids displayed some of the same trends as they got older. When a young organoid first began making waves, they came in sparse bursts. But as the organoid developed over months, they grew longer and better organized, their lulls growing smaller.

This unsettling discovery did not mean that Negraes and her colleagues had created baby brains. For one thing, a human infant's brain is a hundred thousand times bigger than the biggest organoids. For another, the scientists only mimicked one part of the brain—the cerebral cortex. A working human brain has many other parts: a cerebellum, a thalamus, a substantia nigra, and on and on. Some of its parts take in smells. Others handle sight. Still others make sense of different kinds of input. Some parts of the brain encode memories; some jolt it with fear or joy.

Still, the scientists were unsettled. And they had every reason to suspect that, with more research, brain organoids might become more brain-like. A blood supply might let them grow bigger. Researchers might connect a cerebral cortex organoid to a retinal organoid that could sense light. They might link it to motor neurons that could send signals to muscle cells. Muotri even dabbled with the idea of linking an organoid to a robot.

What might happen then? When Muotri started growing organoids, he assumed they could never become conscious. "Now I'm more unsure," he confessed.

So were bioethicists and philosophers. They began gathering to talk about brain organoids and how to think about them. I called one—a Harvard researcher named Jeantine Lunshof—to get her opinion.

Lunshof wasn't too worried about Muotri accidentally creating conscious creatures in a dish. Brain organoids were so small and simple that they still fell far below that threshold. What concerned her was a simple question: What on earth are these things?

"In order to say what you should do with it, you first have to say, 'What is it?'" Lunshof explained to me. "We're making things that were not known ten years ago. They were not in the catalog of philosophers."

In La Jolla, Lunshof's question came to my mind as Trujillo showed me his latest batch of organoids.

"This is just a mass of cells," he said, pointing to one of his wells. "It does not get close to a human brain. But we have the tools to make a more complex mini-brain."

"So you feel okay with this," I said, groping for the right words, "because obviously it's not a human brain—"

"Human cells!" Trujillo clarified.

"So they're alive," I half said, half asked.

"Yes," Trujillo replied. "And they're human."

"But they're not a human being?"

"Yes," he said.

"But where would you start to approach that line?" I asked.

Trujillo had me imagine an organoid rigged up to an electrode. "You can do a pattern of electrical shocks," he said.

Trujillo was sitting in front of a microscope as we talked. He extended two of his fingers and rapped them on the counter, producing galloping beats.

Ba-bap, ba-bap, ba-bap.

He suspended his hand over the counter. "And then we stop."

After a few seconds Trujillo brought down his fingers again. *Ba-bap, ba-bap, ba-bap.*

"And then the thing fires," he said. In response to the incoming signal, the organoid uses its neurons to create a matching signal of its own. "That's a bit more concerning. It's learning something."

We are badly equipped to make sense of these crackling spheres. Our problem is not simply that brain organoids are new. If you get a new smartphone for your birthday, it may take you a little while to figure out how to unlock it, but it doesn't cause a philosophical crisis. Brain organoids are troubling because we feel in our bones that making sense of life should be easy. These clusters of neurons prove that it's not.

To decide whether brain organoids are alive or not, we compare them to the life we know best, the benchmark against which we judge all other possible kinds of life: our own. If someone asks you if you're alive, you don't need to check your pulse or prove to yourself that your cells are breaking down carbohydrates before answering yes. It's just a deeply experienced fact.

"We know what it feels like to be alive," the biologist J. B. S. Haldane observed in 1947, "just as we know what redness, or pain, or effort are." These chunks of knowledge seem powerfully obvious. And yet, Haldane observed, "we cannot describe them in terms of anything else."

People can lose this sense of what it feels like to be alive without actually dying. To the contrary, they insist they are dead. The condition is rare, but people suffer from it often enough that it has earned a name: Cotard's syndrome.

In 1874 the French physician Jules Cotard examined a woman who had been admitted to a hospital after becoming suicidal. He wrote in his note that she "affirms she has no brain, no nerves, no chest, no stomach, no intestines; there's only skin and bones of a decomposing body." The fact that she could express this conviction in fully formed sentences did not sway her from it.

In the generations since, more accounts of Cotard's syndrome have surfaced. In Belgium a woman became convinced her whole body was a translucent husk. She refused to bathe for fear that she would dissolve and vanish down the drain. A man in Germany informed his doctors that he had drowned in a lake the year before. The only reason he could explain his condition to them was that radiation from cell phones had turned him into a zombie.

Because Cotard's syndrome is so rare, neuroscientists have managed to study only a few of the brains of people who experience it. In 2015, Indian doctors described the case of a woman who told her family that cancer had rotted her brain and then claimed her life. An MRI revealed that her skull still contained a working brain. But the doctors noticed that a region a few inches behind her eyes was damaged.

This region, known as the insular cortex, receives signals from across our body. It then generates a conscious awareness of our internal sensations. The insular cortex becomes active when we're thirsty, experience an orgasm, or have an uncomfortably full bladder.

The signals that flow into the insular cortex may be crucial to our intuitive sense of being alive. If it gets damaged, that intuition may abruptly vanish, producing Cotard's syndrome. Our brains constantly update their picture of reality to suit the signals they process. When people no longer get information about their internal state, they update reality to make sense of the change. The only explanation that makes sense is that they're dead.

We don't just know what it feels like to be alive, however; we also recognize life beyond our own skin. For our brains, recognizing other living things is a bigger challenge, since our nerves don't reach into their bodies. We have to bridge the gap with the signals we take in with our sensory neurons—in other words, what we see, hear, smell, taste, and touch.

To speed up this recognition, we use unconscious shortcuts. We take advantage of the fact that living things can direct their own motion toward their own goals. As wolves run down a hillside in pursuit of a moose, they dodge trees and look for ways to cut off their prey. A boulder tumbling down that same hill falls predictably and passively. Our brains are tuned to these differences, recognizing whether an object is showing biological or physical motion in a fraction of a second.

Scientists have found that we can perceive living things so quickly because we only need a tiny amount of information to trigger biological circuits in our brains. In one series of experiments, psychologists filmed people walking, running, and dancing and marked their joints in each video frame with ten dots. They showed movies of these moving dots to people, interspersed with movies of ten dots moving independently of each other. People could quickly tell the difference.

Our perceptions are not the only feature of our brain tuned to life. Our memories are as well. As we build up information about things, we file it in our brains according to whether they are alive or not. Brain damage can expose our filing system. People with damage to certain regions struggle to name living things such as insects and fruits. And yet they have no trouble with toys or tools.

Psychologists have long wondered to what extent we're born making these distinctions and how much we learn them as we grow up. After all, you can immediately recognize the words in this sentence, but that doesn't mean you were born with that skill. Experiments on children suggest that their intuitions about life are present from the start. Infants prefer to look at dots that move in biological patterns rather than random ones. They will look longer at geometrical shapes that seem to be self-propelled than ones that seem to move passively. Children also have a bias toward life in the way they learn: they can learn about animals faster than inanimate objects, and they hold on to the memories of what they learn longer. Our knowledge of life, in other words, arises long before we can tell ourselves what we know.

"If we carve the human mind at its joints," the psychologist James Nairne and his colleagues have written, "the distinction between living and nonliving things forms a natural place to cut."

Our sense of living things is far older than our species. Experiments on animals have revealed they can make some of the same distinctions between the living and the nonliving that we do. In 2006, Giorgio Vallortigara and Lucia Regolin, two Italian psychologists, made a dot movie of their own, but they filmed chickens rather than humans and then showed their movies to newly hatched chicks. If the hen-shaped dots faced the left, the chicks tended to turn that way as well; they tended to turn right if the hen was facing that way. Vallortigara and Regolin didn't observe this behavior when they showed the chicks movies of random dots or if they turned the hen-shaped dots upside down.

Studies such as these suggest that animals have used visual shortcuts for millions of years to recognize other living things. This strategy allowed predators to quickly spot prey. It was good for the prey, too, because it provided crucial information for making a safe escape. Evading a wolf and evading a falling boulder call for two very different—and very quick—reactions.

About 70 million years ago our earliest primate ancestors inherited this ancient instinct for life. But in their subsequent evolution they gained new ways to recognize living things. Their descendants evolved powerful eyes and big brains, with complex networks of neurons that merged their vision with other senses. Along the way, some species of primates became intensely social, often living in large groups. To thrive in a society, they had to become keenly sensitive to the faces of other primates, reading expressions and tracking gazes.

Our ape ancestors arose about 30 million years ago. They evolved even bigger brains, along with a deeper understanding of their fellow apes. Our closest living ape relatives, the chimpanzees and bonobos, can use subtle cues in faces and voices to infer what others are feeling and what they know. They do not have a language to put these inferences into words. Ask a chimpanzee to define life, and you'll be sorely disappointed. Yet an ape still has a deep sense of its fellow apes as living things—the same sense we inherited when our ancestors split off into their own lineage 7 million years ago.

The brain continued to grow in the human lineage; our species has the biggest brain in the animal kingdom relative to our body size. Our ancestors also evolved the capacity for language and an even more powerful ability to get into the heads of other humans. But all these features evolved on top of the foundation we inherited from earlier primates. And that deep foundation may account for our overweening confidence that we know what it means to be alive, even when we don't.

When a new member of our species was born, our ancestors could use their biology-sensing brain circuits to recognize another living human. But they had not evolved any intuition about how that new human life had come about. Instead, people came up with explanations.

In Ecclesiastes, for example, we read about "how the spirit comes to the bones in the womb of a woman with child." Jewish scholars would later teach that the embryo is "mere water" until the fortieth day. Christian theologians combined the Bible with

Greek philosophy to create a different explanation. In the thirteenth century, Thomas Aquinas described a process of "ensoulment." He argued that human embryos first gained a vegetative soul, with the same faculties for growth as plants. The vegetative soul was later replaced by a sentient soul like that of an animal. And later the sentient soul was finally replaced in turn by a rational soul.

Other cultures created their own explanations. The Beng, a group of rural villagers in the Ivory Coast, see the beginning of life as the journey from another world. Babies are spirits from *wrugbe*, a settlement occupied by the dead. Only a few days after birth, when the umbilical cord stump falls off, does a newborn truly belong to this world. If it should die before then, the Beng give it no funeral. There is no death to observe.

Beliefs about how living things get their start gave rise to customs and laws around pregnancy. For ancient Romans a human life began with its first breath. Roman doctors and healers regularly induced abortions in pregnant women by giving them herbs. But a woman had no say in whether she could get an abortion; the decision lay entirely with the patriarch of her family. In medieval Europe, Christian theologians held that fetuses had souls, which meant that abortion was a crime. Yet they debated about exactly what that rule meant for actual pregnancies. Aquinas's followers argued that a distinction had to be made between the early and late stages of pregnancy. In 1315 a theologian named John of Naples gave physicians guidance for cases in which a pregnancy threatened a woman's life. If the fetus was not yet ensouled, the physician should provide the abortion. "Although he impedes the ensoulment of a future fetus, he will not be the cause of death of any man," John declared.

If a fetus had already gained a rational soul, on the other hand, a physician should not try to save the mother's life with an abortion. When "one cannot help one without hurting the other," John wrote, "it is more appropriate to help neither."

The trouble with this kind of guidance was that no one had any idea of exactly when a fetus became ensouled. Some theologians believed the best way for physicians to deal with this uncertainty was to never perform an abortion. Others left the matter up to a physician's conscience. In the sixteenth century, judges in Italy set the threshold for ensoulment at forty days after conception. And in 1765 the British judge William Blackstone came up with a new standard: the quickening.

"Life is the immediate gift of God, a right inherent by nature in every individual," Blackstone wrote, "and it begins in the contemplation of law as soon as an infant is able to stir in the mother's womb."

The American colonies adopted quickening as their standard, too. And for generations abortions were a quiet fact of American life. Pregnant women who sought out abortions suffered little penalty. Housewives medicated themselves with abortion-inducing plants they grew in their gardens. Later, in the industrial revolution, women flocked from farms to cities, where they tried to induce abortions with "female monthly pills" advertised in newspapers. These crude abortion-inducing drugs often failed, forcing the women to find doctors who would surgically carry out the procedure in secret.

Over the course of the nineteenth century, the opposition to abortion grew more organized. Pope Pius IX declared abortion a mortal sin—even before the quickening. In the United States, anti-vice crusaders warned that access to abortions tempted women into sinful lives. The American Medical Association agreed, and prominent doctors gave speeches about the dangers that abortions posed to fetuses and to pregnant women alike. In 1882 a Massachusetts doctor named Charles A. Peabody delivered one such attack, calling on his fellow physicians to resist the pleas of pregnant women for abortions.

"It is a sin against God—a crime of the deepest dye," Peabody warned.

For a doctor like Peabody, educated in late nineteenth-century medicine, the terms of battle over pregnancy were profoundly different from those in earlier centuries. Medieval scholars had little idea of what happened inside a uterus. They relied on the Bible, Aristotle, and a few fetal kicks. Peabody lived at a time when scientists studied sperm, eggs, and fertilization. They tracked the development of embryos. In the late 1800s many scientists still thought of life in terms of a mysterious vital force, the fundamental role of genes and chromosomes still decades away from discovery. Those vital forces were unleashed at the moment of conception.

"When does life begin?" Peabody asked. "Science returns but one answer: no other is possible. Life begins at the beginning, with the first moving of the vital principle, with the first coordination of its forces."

According to this line of reasoning, the law could not use quickening as a line for legal abortions. "No!" Peabody thundered. "Life begins at the beginning, and along the way of his natural journey a human being has a right to his life."

By the time Peabody delivered this attack in 1882, many American states had already passed strict laws banning abortions. Yet loopholes allowed doctors to keep carrying out the procedure as they saw fit. Sometimes they performed abortions for the health of the mothers. Depression, suicide, or extreme poverty could be justification enough. Many doctors were willing to perform abortions for victims of rape. Only rarely did these abortions come to light. And rarer still did a doctor get arrested.

This invisible, semilegal system lumbered along for decades in the United States until a new push against abortions in the 1940s suddenly eliminated many of the safer avenues pregnant women could take. Many got botched abortions, often self-administered, and showed up in hospitals in droves. Hundreds died every year.

Reformers called for a change to the laws. A massive outbreak of measles in the early 1960s produced a wave of devastating birth defects, leading to demands from women for access to safe abortions. States responded by making abortions legal under certain circumstances. In the 1973 case *Roe v. Wade*, the Supreme Court ruled that criminalizing abortion violated a woman's right to privacy. States could restrict abortions only after the second trimester, they ruled, once a fetus became viable to survive outside of the womb.

In their decision, the court addressed the start of life—only to say they did not have to address it. "We need not resolve the difficult question of when life begins," the court declared. "When those trained in the respective disciplines of medicine, philosophy and theology are unable to arrive at any consensus, the judiciary, at this point in the development of man's knowledge, is not in a position to speculate as to the answer."

Antiabortion groups responded to the court's ruling by searching for a way to block abortions that didn't clash with the decision. They boycotted companies that did research into abortion drugs. They lobbied for laws making it hard for abortion clinics to do their work. To win over voters, they invoked new scientific research—or at least a carefully selected version of it.

They claimed that studies on fetuses pushed back the time at which they began to feel pain. Some antiabortion legislators introduced "fetal heartbeat" bills. They skipped over the fact that hearts do not yet exist when cardiac cells start to contract. The bills didn't have anything to do with actual hearts anyway, since their purpose was to effectively ban most abortions after just six weeks.

Beyond these half measures, many antiabortion groups wanted to overturn *Roe* altogether. The only way to do that was to address the question of when life begins—or, to be legally precise, to decide when an embryo becomes a person, with all the rights that come with personhood. A so-called personhood movement arose, claiming that these rights extend back to fertilized eggs. If they did, those rights would make any abortion illegal.

Some leaders of the personhood movement acknowledged that certain forms of contraception would also have to be banned, because they blocked pregnancies by preventing newly formed embryos from implanting in the uterus. And to justify this legal case, they invoked science in much the same way Charles Peabody had over a century earlier.

"Life begins at conception," the conservative pundit Ben Shapiro declared in 2017. "That's not religious belief. That's science."

Shapiro, it should be pointed out, was not a scientist. He had a law degree and a podcast. And when he made this claim, he did not offer scientific evidence to back it. Scientists, on the other hand, have been pushing against these sharp, all-or-nothing claims about life ever since the molecular underpinnings of life became clear. In 1967, in the pre-*Roe* battles over abortion, the Nobel Prize–winning biologist Joshua Lederberg addressed the controversy with a piece called "The Legal Start of Life" in the *Washington Post.*

"There is no single, simple answer to 'When does life begin?'" Lederberg wrote. "In contemporary experience, life in fact never begins."

A fertilized egg is alive, Lederberg explained, but in the way cells are alive, not people. Some organisms, like bacteria, spend their entire existence as single cells, thriving happily in the ocean or the soil, but the cells that make up our bodies are not so rugged. If you prick your finger and dab a drop of blood on a table, your cells will not crawl off to seek their fortune. They will dry out and die. For cells, death means that their proteins malfunction, their interiors are thrown off chemical balance, and their membranes tear open. Inside a body, a cell can thrive. It can feed on the nutrients that wash over it, keep its proteins in good working order, and get rid of its waste. If it gets the right signals, it can grow and divide. One cell becomes two, as the so-called mother cell splits up all its molecular legacy between a pair of new daughter cells. At no point during cell division does the mother cell die. At no point do the daughter cells come to life. What gives life flows from the former to the latter.

Some types of cells can run this movie backwards. Instead of dividing, they fuse together. When we exercise, for instance, we stimulate muscle cells to multiply and then merge in order to create new fibers. In our bones, immune cells fuse into giant blobs called osteoclasts, which nibble away at old bone so that it can be replaced with new tissue. Each muscle cell and osteoclast can hold many nuclei, each packed with its own DNA. The independent cells that came together to form them did not die. They simply mixed their molecules together into a new form of life.

This is the cellular universe in which a fertilized egg exists. It is certainly alive, but it does not snap into life thanks to the assembly of lifeless molecules. Instead, it emerges from the fusion of two living cells. But the mother's egg and the father's sperm from which it arose did not jump into existence, either. The egg arose from cells that divided when the mother was still an embryo. A man makes hundreds of millions of sperm each day, but ultimately they all descend from the fertilized egg that gave rise to his entire body. The flow of life arrives unbroken from the previous generation, and from generations back through the ages. You'd have to canoe up life's river for billions of years before reaching its headwaters.

"Life begins at conception" is a simple slogan, easy to remember, easy to shout. Taken literally, though, it's false on its face. The personhood movement's politics have always made it plain that the slogan wasn't supposed to be taken literally anyway. It's not *life* that they're talking about starting at conception but *a life*. And not just any life—not the life of an armadillo or a petunia—but a human life, with all the legal protections it is due, including—to close the circle—the right to life.

"A distinct, living human individual comes to be with the fertilization of the oocyte by the spermatozoan," Patrick Lee and Robert George, two abortion opponents, wrote in 2001. What makes it distinct, they argued, is that it has a unique set of DNA, combined from its two parents, that can guide its development. It might be invisible to the naked eye, but Lee and George argued that the fertilized egg already has the potential for reasoning and all the other capacities that make us human.

The actual course of human development makes it impossible to pin one instant as marking the origin of a new human individual. It certainly can't be the moment that a sperm fuses with an egg. Cells typically carry forty-six chromosomes—twenty-three from our mother and twenty-three from our father. But at the moment of fertilization, the combination of a father's and mother's DNA actually produces sixty-nine chromosomes. That's because an unfertilized egg is a cell like any other cell in a woman's body, with forty-six chromosomes arranged in twenty-three pairs.

A cell with sixty-nine chromosomes could never give rise to a healthy human being. Its genes would be wildly out of balance. To avoid this catastrophe, an egg responds to the arrival of a sperm by pinching off a tiny bubble. Inside that bubble, the egg stows away twenty-three of its chromosomes. The egg is now left with the other twenty-three—a perfect counterpart to the father's DNA.

Even now, however, the fertilized egg has not gained a single new genome we can call its own. Its mother's and father's chromosomes still remain separate, swaddled in their own membranes, in which they undergo separate changes. It's better to think of the early fertilized egg as a coworking space, a place in which the male and female genomes busy themselves on their own.

The fertilized egg then divides into two cells, each of which inherits the chromosomes of both the father and the mother. It takes a day after fertilization to reach this milestone. And only then do the chromosomes abandon their separate containers. Only in the two-cell embryo do the two sets of DNA join together.

And yet even at this point the new embryo does not have its molecular independence. Virtually all the proteins in the cells come from the mother, encoded by her genes. In this important respect, the embryo still behaves as if it were a cluster of the mother's cells. A distinct, human individual is not yet taking hold of its own fate. Before the father's chromosomes can wake up—before the new genome can take charge—there's a lot of work yet to be done. Inside the egg are a special set of assassin proteins, made from the mother's own genes. They roam the embryo's cells, annihilating her other proteins. Another set of her proteins grabs hold of both her and the father's chromosomes and prepares them for their new job. Now the cells make a fresh batch of proteins, rebuilt from the shredded remains of the mother's molecules.

As these changes take place within an embryo, it floats out of the mother's oviduct and down into her uterus. Along the way it may break in two. The two clusters of cells continue to divide, each becoming an ordinary embryo. Ultimately, these two sets of cells can develop into identical twins. If we must believe that a fertilized egg immediately becomes a person, then we're left to wonder where that person went when it became two people.

Fraternal twins develop in a different way. The mother releases two eggs at once, each of which is fertilized by a different sperm cell. Sometimes, when these twins are still tiny clumps of cells, they bump into each other and merge. Thanks to their flexibility, the cells reorganize themselves into a single embryo that continues to develop normally, even though some cells contain one genome and the other cells contain another.

Scientists call these mergings chimeras. Chimeras can grow into healthy adults who go through life made up of two populations of cells, each with their distinct genome. If every fertilized egg is a single person with all the rights that a single person is entitled to, does a chimera get to have two votes?

When we living humans look back at the development of an embryo, it's tempting to see it as a gorgeously precise clockwork of chemistry that transforms a single cell into a 37-trillion-cell body. Textbooks illustrate every stage proceeding without a glitch. But development often ends in failure, with many pregnancies lost along the way. The biggest risk to the survival of an embryo is if it doesn't end up with twenty-three pairs of chromosomes. Sometimes it ends up with a third copy of a chromosome. With three copies of each gene instead of two, an embryo may make too many proteins, poisoning itself. Embryos may end up with just one copy of a chromosome, leaving them unable to make all the proteins they need to survive.

Sometimes the imbalance arises in the egg. When the egg tries to get rid of its extra chromosomes in a bubble, one of them accidentally stays behind. Other times the trouble comes after fertilization, when the embryo starts to divide. As the cells split, they may fail to divide their chromosomes equally between their daughter cells. One cell may end up with too many chromosomes and the other with too few. As they divide, they pass down that imbalance to their descendants.

Biologists call this imbalance aneuploidy. It doesn't necessarily spell doom for an embryo. If it contains balanced and unbalanced cells, the unbalanced ones may stop growing, while the balanced ones go on to make up the vast majority of the body. Even if an embryo is made up entirely of aneuploid cells, it may still have a chance of surviving. It depends on the nature of the imbalance. An embryo with an extra copy of chromosome 21 may be born as a child with Down syndrome. In most cases, however, aneuploid embryos fail. Sometimes they simply stop growing. Sometimes they fail to implant in the uterus and get flushed out.

Aneuploidy is not the only cause of lost pregnancies. Some women can't make enough hormones needed to prepare their uterus to take in a new embryo. A badly timed infection may overcharge a woman's immune system, which then treats embryos and placentas as foreign enemies to be attacked.

Scientists have come up with estimates for how many pregnancies are lost naturally, and they're enormous. One study published in 2016 concluded that between 10 and 40 percent of embryos are lost before they can implant in the uterus. All told, from conception to birth, the researchers found that the figure may rise to 40 to 60 percent. If a country were to declare that life begins at conception, and that fertilized eggs have the legal rights that all persons are due, it would have to treat these losses as a medical catastrophe. Worldwide, it would mean the deaths of perhaps more than 100 million human beings every year, dwarfing the deaths from heart disease, cancer, and every other leading cause.

Yet this crisis hasn't become an urgent priority for opponents of abortion. Just the opposite: some of them have questioned these estimates, suggesting the losses are somewhat smaller—as if tens of millions of deaths would somehow be easier to live with. Some claim that the causes of these lost pregnancies, such as aneuploidy, are unstoppable, so these lives couldn't be saved anyway. But that's not true. A great deal of research has gone into reducing pregnancy losses—not because researchers subscribe to the idea that life begins at conception, but because they want to help couples struggling to have children. Some women who have recurrent pregnancy losses can improve their odds of a successful birth by getting hormone injections. Other researchers are exploring new possibilities for saving embryos, from managing a mother's immune system to editing the DNA of fetal cells.

Abortion opponents also undermine their own sweeping claims with illogical exceptions. In 2019, Alabama legislators introduced a bill that would charge doctors who carried out abortions with a felony. They would face a punishment of up to ninety-nine years in jail. But the bill's authors made an exception for women who faced serious health risks from their pregnancy. When the bill attracted controversy, the Alabama Senate Judiciary Committee tacked on additional exceptions, for rape and incest.

One of the bill's sponsors, state senator Clyde Chambliss, objected. "In the situations of rape and incest it is a very difficult, difficult situation following a horrendous act," Chambliss told reporters. "But if we believe that life begins at conception, and I do, then life is lost."

But Chambliss couldn't follow his own rule to its logical end. When couples use in vitro fertilization to have children, fertility doctors routinely make a batch of embryos, not just one. They may pluck one cell from the embryo to closely examine its DNA to see how viable the embryo will be. Since all the cells in early embryos can become embryos of their own, this test should, by Chambliss's logic, cause the loss of life. Once fertility doctors choose the best embryos for implantation, they may freeze or discard the others. If abortions are unjustifiable because embryos are persons, then it is unjustifiable to let embryos die as a result of in vitro fertilization. It doesn't matter whether they are actively or passively killed.

Yet, during the debates over the Alabama bill, Chambliss declared his ban did not stop in vitro fertilization. When a fellow lawmaker challenged him on this inconsistency, he gave an inscrutable response.

"The egg in the lab doesn't apply," he declared. "It's not in a woman. She's not pregnant."

The Alabama legislature went on to vote down the amendment to allow abortions in the case of rape and incest. The governor signed the bill.

In vitro fertilization complicated the question of life's beginning, and now reprogrammed cells promise to complicate it even more. With the right combination of chemicals, a reprogrammed cell can start developing into an embryo. Scientists have turned the skin cells of adult mice into mouse embryos, which can grow into mouse pups. It may soon be possible to do the same with humans. When that happens, trillions of cells in each of our bodies will gain the potential to become a human being. According to the logic of the personhood movement, they will all be due the rights of a person. The dust in our homes is largely made up of the dead skin cells we slough off by the millions every day. Is each one a potential life lost?

None of these complications means that we can walk away from our moral obligations to our fellow humans. It just means there's no easy way to figure them out. And as organoids become more complex, it may get especially hard to decide what our moral obligations are to them. Today's brain organoids are alive, yes, and they are human, but they don't experience the life that human beings do. That life has something to do with Haldane's feeling of being alive. It's conceivable that a bigger, more complicated organoid might make intricate brain waves, might even learn. Perhaps it might even gain a rudimentary sense of life.

How could we find out if it gained that sense? Christof Koch, the director of the Allen Institute for Brain Science in Seattle, has an idea. He thinks scientists could measure the complexity of an organoid's experiences by eavesdropping on its signals. Koch's proposal emerges out of work that he and other scientists have done on the nature of consciousness. They argue that consciousness is the integration of information across the brain. When we are conscious, information flows across our whole brain, giving us a coherent feeling of reality. When we fall asleep or go into a coma, the flow dwindles down. The regions of the brain remain active, but their information no longer adds up to a single, unified experience.

Koch and his colleagues believe that we can measure this integration by disturbing it, like tossing a rock into a pond to look for ripples. They've put magnets on the heads of volunteers and delivered harmless pulses. The pulses briefly disturb their brain waves. In people who are awake, the pulses produce flows of information traveling along complex paths through the brain. The same pattern arises when people dream. But when people go under anesthesia, the pulses trigger simple responses—like the ringing of a bell instead of a fugue played on a pipe organ.

Koch has suggested that scientists could apply the same magnetic pulses to brain organoids and see how they respond. What makes his proposal particularly intriguing is that he and his colleagues have invented a way to measure the integration in a brain with a single number. It's like a thermometer for consciousness. We might agree that brain organoids should never rise above a certain number. And if we discovered that a particular batch of organoids managed to sneak past the threshold, we'd know that we have to decide how we'll care for their lives.

"What would it mean for a cerebral organoid to suffer?" Koch asked at the end of a lecture he gave in 2019. "That's not an obvious question to answer."

In 1967, long before the dream of organoids even existed, Joshua Lederberg could see the trouble that lay ahead.

"The biologist, then, is not really very helpful to the law," Lederberg said. "The question of when life begins is answered according to the purposes for which we ask it."

DEATH IS RESISTED

In 1765 a fifteen-year-old boy named James Forbes boarded a ship in England and sailed for Bombay. There he joined the East India Company, and over the next nineteen years his job took him back and forth across the subcontinent. Along the way, Forbes turned himself into a naturalist and artist, painting portraits of bulbul birds and Parsee families. By the time Forbes left India to return to Europe, he had produced 52,000 pages of writing and art.

Back home he combed through his work, and in 1813 he published a four-volume book called *Oriental Memoirs* offering a sumptuous tour of India for his British fireside readers. The *Monthly Magazine* praised "the TRULY SPLENDID work before us." With his encyclopedic scope, the editors believed, Forbes made a visit to India pointless. "He has left little of novelty to be discovered by future travelers."

Along Forbes's journeys, he stopped off at a great banyan tree on the banks of the Narmada River. It sent hundreds of trunks into the sky, creating a canopy big enough to shelter an army of seven thousand soldiers. A local chief sometimes visited the tree to host giant parties. He set up lavish tents that served as a dining room, a drawing room, a saloon, a kitchen, and bathrooms. He had enough free space left over to fit his camels, horses, carriages, guards, and attendants—along with his friends and their herds of cattle.

The Narmada banyan was also home to birds, snakes, and langur monkeys. Forbes observed the monkeys teaching their young how to leap from tree to tree and kill dangerous snakes. "When convinced that the venomous fangs are destroyed, they toss the reptile to their young ones to play with, and seem to rejoice in the destruction of a common enemy," Forbes said.

A friend of Forbes once paid a visit to the Narmada banyan as part of a shooting party. He shot a female monkey with his fowling piece and took the corpse to his tent. A cacophony of screeching began outside the tent walls, and when he looked out, the hunter saw dozens of monkeys "who made a great noise, and in a menacing posture advanced towards it," Forbes said.

Forbes's friend brandished his fowling piece. The animals fell back, with the exception of one male, who seemed to be the leader of the troop. The monkey approached the hunter, chattering aggressively. But eventually his calls changed to what Forbes described as "a lamentable moaning."

It seemed to the hunter that the monkey was begging for the dead female's body. He gave it back.

"With tender sorrow he took it up in his arms, embraced it with conjugal affection, and carried it off with a sort of triumph to his expecting comrades," Forbes wrote. After the monkeys departed, the entire shooting party was left shaken. "They resolved never more to level a gun at one of the monkey race."

Forbes's story of the lamentably moaning monkey was so remarkable that people in England repeated it for decades. It seemed to fly in the face of what Victorians thought about the animal brain. Humans could make sense of life, thanks to their rational minds. And through understanding life, they could also see its limit in death. But here were brutes that acted remarkably like humans in mourning, that seemed to know that the life in their fellow monkey had gone. One might conclude that monkeys had more sophisticated minds than we gave them credit for. Or perhaps we humans flatter ourselves too much about what we know about life and death.

Forbes's story of the mourning monkey was joined by other stories of grieving primates, and no one was more fascinated by them than Charles Darwin. Once he conceived of his theory of evolution in his late twenties, Darwin recognized that it explained the origins of humans just as it did any other species. He could see the legacy of evolution in our anatomy, with its striking similarity to chimpanzees and other apes. He paid visits to an orangutan at the London Zoo and could see the legacy in her humanlike facial expressions. And he could see it in the stories of primates displaying emotions that were once thought unique to our own species. Among them were tales of grief. "So intense is the grief of female monkeys for the loss of their young, that it invariably caused the death of certain kinds," Darwin wrote in his 1871 book *The Descent of Man*.

Nearly a century would pass before scientists regularly traveled to the wild habitats of monkeys and apes to make detailed observations of their behavior. But once they got there, they began accumulating their own firsthand stories of the striking ways that primates responded to death. Researchers came to see these stories as a scientific question in its own right, which they called primate thanatology. The first modern record of primates facing death came in the 1960s from Jane Goodall, a young British naturalist who traveled to Tanzania to live with chimpanzees. One day Goodall dedicated her observations to a female she called

Olly. Olly had recently given birth, but Goodall could tell the baby was not well. "All his four limbs hung limply down," she later recalled, "and he screamed almost every time his mother took a step."

Since the baby was too weak to grip Olly's hair, she had to carefully cradle him. She took him up a tree, where she sat on a branch and carefully placed him in her lap. A blinding rainstorm swept in and drenched chimpanzees and primatologist alike for half an hour. When it cleared away, Goodall watched Olly climb down to the ground again. The baby now made no sound. His head hung as lifelessly from his body as his limbs. And now Goodall noticed that Olly treated her baby differently.

"It was as though she knew he was dead," Goodall said.

Rather than cradling the infant, Olly now held him by a leg or an arm. Sometimes she slung his body around her neck. In a seeming daze, she carried her baby for the next two days. Other chimpanzees gawked at her and her dead infant, but Olly simply stared off into space. Eventually, Goodall lost track of her as she traveled through a dense thicket. She did not catch up with Olly until the following day. The baby was gone.

In the decades since, other primatologists have seen other mothers respond to the loss of their infants much as Olly did. They have observed young gorillas sitting vigil with their dead mothers. While working in the forests of the Ivory Coast, Christophe Boesch once came across the body of a chimpanzee on the ground. It looked as if it had just died after falling out of a tree. He then saw five other chimpanzees arrive and spot the body, too. They swiftly climbed into the surrounding canopy, where they hooted and screamed for hours.

Our sense of what it means to be alive emerges partly from our awareness of our own life, and partly from our intuitive ability to tell living things apart from inanimate objects. But it also grows out of our understanding of the difference between life and death. To be alive is to not be dead, in other words. Humanity did not come to this realization through logic and deduction. Our understanding of death is not like Darwin's theory of evolution or Thomson's discovery of the electron. It has its origins in ancient intuitions.

Animals probably first evolved to behave differently toward living and dead things hundreds of millions of years ago. Today, mammals, birds, and even fish can be put off by the smell of rotting bodies. The disgusting smell of death is the result of certain airborne molecules with evocative names like *cadaverine* and *putrescine.* These molecules are not produced by death, however, but by life growing on death. After an animal dies, its cells self-destruct and become food for the body's resident bacteria. They chew through the walls of the gut and spread through the body. They release cadaverine and putrescine merely as byproducts of their metabolism. These molecules are not actually dangerous to us. They won't kill us like a whiff of sarin or cyanide. Yet our ancestors evolved a keen sensitivity to these molecules, along with an instinctive response to recoil at the merest whiff. That's because they are reliable signals of the dangers of the dead, even if they're not dangerous themselves.

Thanks to primate thanatology, we now know that our monkey-like ancestors 70 million years ago did not have to wait for their dead compatriots to start rotting to sense that something important had happened to them. This keener sense of death, some scientists have argued, resulted from a keener sense of life. When a primate died, living primates around it would still see features like the eyes and mouth that triggered their biological detection circuits. But the circuits dedicated to biological motions would register nothing—not even a blink. These contradicting signals may explain why primates so often sit vigil with their dead. They may need time to make sense of this cognitive clash, to move a primate that they've lived with for years to the category of the lifeless.

By about 30 million years ago, the lineage of apes—the evolutionary branch that would produce orangutans, gorillas, chimpanzees, and us—split off from other primates. Studies on chimpanzees suggest that our common ape ancestors evolved an even deeper sense of death, perhaps as a result of evolving bigger, more powerful brains. Chimpanzees not only react differently to fellow chimpanzees when they die: they also show some signs of recognizing the cause and effect of life and death. Their behavior suggests that they comprehend that a fall out of a tree or the attack of a leopard can bring the life of an ape to an end.

Our own lineage split off from that of chimpanzees roughly seven million years ago. Early hominins gradually evolved to walk upright in woodlands; aside from that they didn't look much different from other apes. Nor is there a sign in the fossil record that they treated their dead any differently than apes do. Only in the past few hundred thousand years do the first hints of a modern sense of death appear. And the oldest of these hints, not surprisingly, are the most ambiguous.

In a few caves in Africa and Europe, paleoanthropologists have found caches of early human skeletons. These humans belonged to our own genus, *Homo*, but to two different species: *Homo heidelbergensis* and *Homo naledi*. It's conceivable that the skeletons of these early humans were ceremoniously carried to their resting places and then dropped down fissures. But for now the evidence is still too patchy to be sure. It's also possible that predators dragged these early humans into caves or roaring floodwaters washed them in.

The oldest indisputable evidence of a new concept of death dates back about 100,000 years. Members of our species, *Homo sapiens,* began carrying out funerals. In caves in Israel, archaeologists have found skeletons that were carefully laid out and surrounded with deer antlers, chunks of ochre, and shells from distant shores. In Australia, Aboriginals were digging graves for their dead by about 40,000 years ago. These rituals tell us something about the minds of the people who practiced them. They were understanding death in a way that other primates did not: that diseases and injuries were its cause, that there was no going back. They honored the memory of the dead by carefully interring their bodies.

By the time people were performing these first funerals, they were capable of full-blown language. The echoes of the songs they may have sung or the stories they may have told dispersed long ago. To trace the origin of our concept of death, we have to content ourselves with written accounts and the spoken words of people recorded from across the world. It's clear that humans have come up with many explanations of death, but they share some things in common. People don't simply think about it as a physical change. They also see death as a social transformation. Some cultures conceive death as a separation, as the deceased travel to another world. Others see death as a transformation that enables their ancestors to be always with them. Buddhists, meanwhile, think of it as the disappearance of the self, like the dewdrop on a blade of grass evaporating into the air at dawn.

Western science was slow to create a detailed account of death. It was mostly left to physicians, who were too busy trying to save lives to explain what they were trying to stave off. "Medical men have rarely discussed the so-called sense and essence of death; they had to leave this to philosophers and theologians," the historian Erwin Ackerknecht once wrote.

The first physician to investigate the nature of death in a scientific manner was arguably the French doctor Xavier Bichat. In the late 1700s, he studied both humans and animals in the moments after death. After criminals were executed at the guillotine, Bichat examined their severed heads and headless bodies. He slit open the chests of living dogs to fit stopcocks on their windpipes. With a twist, he could close the flow of air into a dog's lungs. After the dog's blood turned from red to black, he discovered, death was not far off.

This grisly work let Bichat see an intimate connection between the heart, the lungs, and the brain—a vital tripod, as it came to be known. If the lungs failed, they could not transform dark blood to red, the life-sustaining form that the brain needed to keep working. If the heart failed, it could not deliver blood to the other two organs. When Bichat damaged the brains of animals, he discovered that a crucial connection to the heart and lungs was lost, causing the animals to die. No one part of the body had a monopoly on the forces of life, Bichat could see. Those forces were distributed across the body in an interconnected system.

"Life," Bichat concluded, "consists in the sum of the functions, by which death is resisted."

Bichat saw a gleaming line dividing life and death, but its brightness was the result of the kinds of life he studied. Decapitated criminals and exsanguinated dogs left little doubt which side of the line they were on. If Bichat had studied other animals, however, he would have encountered a blurry boundary.

In the late 1600s a Dutch trader named Antonie van Leeuwenhoek crafted the first microscopes powerful enough to open up the microscopic world. A single drop of water from a pond might contain a swarm of strange shapes. They looked unlike anything in the macroscopic world, but they moved, and their movements touched Van Leeuwenhoek's instinctive sense of what makes things alive. He thought of them as little animals. When his reports appeared in the *Philosophical Transactions of the Royal Society*, his English translators used the word *animalcules*. "The motion of most of these animalcules in the water was so swift, and so various, upwards, downwards, and round about, that 'twas wonderful to see," he reported.

Van Leeuwenhoek went on to discover red blood cells, sperm cells, bacteria, protozoans, and a host of miniature animal species. And then, on a summer day in 1701, he noticed that the lead gutter hanging off the front of his house was full of reddish water. He scooped up some of it and put a drop in his microscope. He now saw a new kind of animalcule. These creatures were shaped like pears, with what looked like two wheels atop their heads. (Today they're known as rotifers, meaning "wheel-bearers" in Latin.)

Van Leeuwenhoek then let some of the gutter water evaporate. He had tried this experiment on other animalcules before, and usually they burst as they dried. But this time was weirdly different. As a rotifer dried out, it shrank into a smaller version of itself and became motionless. "It preserved its oval and round shape unhurt," Van Leeuwenhoek observed.

As the summer grew hot and dry, the red water in Van Leeuwenhoek's gutter turned to dust. He decided to search the dust for rotifers, dousing it with water and looking at the drops with his microscope. He spotted more shrunken rotifers, lying in a motionless heap as if they were dead. But after they soaked for a while, they expanded, and then they started to move.

"In a short time afterwards they began to extend their bodies, and in half an hour at least a hundred of them were swimming about the glass," he later wrote.

Van Leeuwenhoek stored away the rest of the gutter dust. Months later he took it out again and mixed it with water. The rotifers unfolded their bodies and came alive even after all the time that had passed.

"I confess I never thought that there could be any living creature in a substance so dried as this was," he said.

Four decades later, in 1743, a British naturalist named John Needham discovered another creature capable of resurrection. Needham had been studying stalks of wheat that were sick with earcockle disease, which caused their grains to swell and turn black. Farmers called these sickened grains peppercorns. When Needham cut open a peppercorn, he found a clump of dry white fibers inside. He added a drop of water to them, hoping it would become easier to pull them apart.

The *Philosophical Transactions of the Royal Society of London* described what happened next. "To his great surprise, these imaginary fibres, as it were, instantly separated from each other, took life, moved irregularly, not with a progressive, but twisting motion; and so continued for the space of 9 or 10 hours, when he threw them away."

Needham had discovered the larvae of a nematode worm, today known as *Anguina tritici*. But at the time, many naturalists refused to believe him. The Royal Society handed Needham's wheat over to another naturalist named Henry Baker to judge.

Baker tried out Needham's instructions and brought the worms to life. His curiosity now aroused, Baker carried out more experiments of his own. In one study he stored some peppercorns for four years. The worms survived over that long stretch of time; when he added water to their white fibers, he saw more writhing life.

"We find an Instance here, that *Life* may be suspended and seemingly destroyed," Baker announced in his 1753 book, *Employment for the Microscope*. How the worms could hold on to their "*living Power*," as Baker called it, he would not dare guess. "What Life *really is*, seems as much too subtile for our Understanding to conceive or define, as for our Senses to discern and examine."

Soon a third animal joined the undead ranks of the nematodes and rotifers. Tardigrades, which look like headless eight-legged bears, only get to be as big as the period at the end of this sentence. Naturalists first discovered tardigrades crawling over mats of moss and later found them lurking in damp soil, in lakes, and even in the ocean. When researchers let tardigrades dry out, the animals retracted their legs and their bodies took on the appearance of sesame seeds. A few minutes of water was enough to sprout their legs again.

Many naturalists refused to believe life could survive this desiccated limbo. They believed something simpler must be happening. Perhaps the desiccated animals died, and when scientists applied water, they awoke hidden eggs that hatched. The battle raged for decades, the two sides coming to be known as the resurrectionists and the anti-resurrectionists. The debate became so dire that France's leading organization of biologists, la Société de Biologie, appointed a special commission in 1859 to settle the matter. After spending a year running experiments, the august scientists issued a 140-page report that came down in favor of the resurrectionists. Yet the anti-resurrectionists kept battling against their conclusion for decades.

Today all biologists are resurrectionists. There is no doubt that tardigrades, nematodes, and rotifers can dry up and then come back to life. The more that researchers study these animals, the longer it seems that they can survive in limbo and still be able to return to the world of the living. In the 1950s a team of researchers collected dried tardigrades in Antarctica. They put the animals in cold storage for thirty years, after which water and warmth brought them back as healthy as ever. The nematodes that occupy peppercorns have lasted even longer, returning from lifeless fibers after thirty-two years.

In recent decades, scientists have added flies, fungi, bacteria, and other species to the ranks of the resurrectable. In Antarctica, glaciers retreated to reveal moss that had been dried out and frozen for at least six hundred years. With some tender gardening, it produced new green sprouts. In Siberia, scientists came across burrows dug by Ice Age squirrels 30,000 years ago that contained dried-out bits of a flower called narrow-leafed campion. The scientists nurtured the fragments, which grew into healthy new plants that produced seeds of their own.

What today's resurrectionists have yet to figure out is how these creatures survive this transformation. In ordinary species, water is essential to a flurry of chemical reactions that take place in every cell every second. Water also helps keep membranes in their proper oily consistency, and it cradles proteins so that their arms and sheets stay in their proper arrangement. When a cell loses water, chemical reactions grind to a halt. Its proteins stick to each other and form toxic clumps, while its membranes turn to sticky jelly. Our bodies can endure the temporary loss of a little water—the kidneys make less urine; the heart beats faster to increase the delivery of oxygen to cells—but once we lose more than a few percent of our body weight in water, our organs start to fail and death soon follows.

Tardigrades and their ilk, on the other hand, can lose all their water. You could argue that they are no longer alive, since they cannot carry out the chemical reactions required for life. Yet neither are they dead. If you pour water on someone who has just died from dehydration, they will not sit up. You are left with a damp corpse. But if you pour water on a dried-out tardigrade, in a matter of minutes it becomes a moving, feeding, reproducing animal. This gray zone of existence has earned a name of its own: *cryptobiosis*—what one team of scientists has described as "a third state between life and death."

When a cryptobiotic species starts to dry out, it eases its own passage into limbo. Some species respond to dehydration by making a sugar known as trehalose. Thanks to its chemical structure, trehalose can help proteins keep their proper shape, much like water does. But unlike water, it doesn't evaporate in dry conditions. This supply of fake water buys the drying creature extra time to make preparation for a long spell of cryptobiosis. Many species make a new batch of proteins that link together to form a kind of biological glass. It entombs the cell's DNA and other molecules in their three-dimensional form, so that they're ready to revive when water returns.

This third state is so durable that it doesn't just give organisms the power to resist dehydration. They can also survive in outer space.

In 2007 a team of scientists collected tardigrades in Germany and Sweden, dried them out, and loaded them into a canister. The canister was placed in a Russian rocket that went into orbit around Earth. For ten days the animals were exposed directly to the vacuum of space. Back on Earth, a splash of water resurrected them.

In 2019 humans delivered tardigrades far deeper into space. An organization called the Arch Mission Foundation set out to create what its founder described to *Wired* as "the hard backup of this planet." They created a miniature "lunar library" in which they stored 30 million pages of information, along with samples of human DNA and thousands of dried-out tardigrades. A private Israeli aerospace company put the library on the Beresheet lunar lander, which they launched to the moon.

The engine failed just before landing, and the Israeli engineers lost track of the probe. It almost certainly crashed into the moon. It's possible that the library sits at the impact site unharmed. The earth rises and sets over the tardigrades as they wait, their cells locked in a glass tomb between life and death, for water that will never come.

As Van Leeuwenhoek put his little animals into a deathlike state, people across Europe were worrying that they might slip into one of their own. They read pamphlets full of terrifying tales of seizures that left their victims without breath or heartbeat. Mistaken for dead, they were lowered into graves, waking up in their coffins when it was too late to be saved.

The fear of this Gothic terror gained strength throughout the eighteenth century and only grew more terrifying in the nineteenth. Edgar Allan Poe mined the nightmare for his story "The Premature Burial," which he published in 1844. "The boundaries which divide Life from Death are at best shadowy and vague,"

Poe wrote. "Who shall say where the one ends, and where the other begins?"

Families made frantic by these stories bought coffins equipped with a string and a bell, so that their not-quite-departed loved ones could sound the alarm. In the 1800s, many German cities built ornate "waiting mortuaries" where the apparently dead could be housed until they began to rot. Mark Twain visited one of these establishments on a trip to Munich in the early 1880s.

"It was a grisly place," he later wrote. "Along the sides of the room were deep alcoves, like bay windows, and in each of these lay several marble-visaged babes, utterly hidden and buried under banks of fresh flowers, all but their faces and crossed hands. Around a finger of each of these fifty still forms, both great and small, was a ring; and from the ring a wire led to the ceiling, and thence to a bell in a watch-room."

It was all an elaborate waste of time: the fears of premature burial were fueled by rumor rather than evidence. But without a quick, foolproof way to determine death, doctors couldn't calm uncertain next of kin. One physician recommended giving patients an enema of tobacco smoke. If they didn't react, they could be safely declared dead. By the mid-1800s a number of doctors were adopting the newly invented stethoscope. Even a faint *lub-dub* meant patients were still alive. Only a long silence came to be the reliable sign that people were truly gone.

Bichat had recognized why a stopped heart is a good sign of death. It belongs to the vital tripod, along with the brain and lungs. If the heart fails, the other two will fail as well. In the twentieth century, scientists mapped these failures in cellular detail. The heart may fail if it can't get enough oxygen from lungs that are scarred or filled with fluid. The cells of the heart need oxygen and sugar to make fuel, and without fuel they cannot contract. If they cannot contract, then the heart cannot send blood to the brain. Brain cells are even hungrier for oxygen than heart cells, and within minutes they will start dying.

A blow to the head can also stop the heart. The impact causes the brain to crash into the interior wall of the skull, ripping apart delicate blood vessels. As the blood gushes in, the brain swells and squeezes its way to the back of the head and then down to the opening at the base of the skull. The pressure shuts off blood vessels throughout the brain, cutting off the supply of oxygen to wide swaths of tissue. The brain stem—the region of the brain that sends out signals required by the heart to beat as well as the lungs to breathe—is often the first part to die.

Bichat was right to believe that, by understanding death, doctors would be better able to protect life. They learned how to treat the loss of blood with transfusions. They learned how to block poisons and fight pathogens. In the early 1900s, American doctors faced a wave of polio outbreaks that left thousands of children paralyzed and slowly suffocating to death. Engineers developed iron lungs to breathe for those young patients. The iron lungs used pumps to create negative pressure around the bodies of the children, drawing air into their lungs. In effect, they were propping up the vital tripod long enough for the children to fight off the virus and regain their ability to breathe on their own.

By the 1950s, iron lungs had given way to tubes that could push air directly into a patient's airway. The advent of polio vaccines made epidemics of paralysis a thing of the past, but doctors still used artificial ventilation on other patients: on victims of drug overdoses, people who had fallen into icy lakes, premature babies—anyone who needed help breathing as they regained good health.

The French neurologists Pierre Mollaret and Maurice Goulon

came to see ventilators as a mixed blessing. They saved many lives but dragged out the ends of others. When people suffered massive brain damage, ventilators could keep their hearts and lungs working, but their brains would never recover. Mollaret and Goulon took careful notes on the outcomes of these patients and found that even with the help of ventilators they didn't wake up again. Instead, they usually died within hours or days. All the ventilators seemed to do was draw out the pain of their families.

This futile condition was, Goulon once said, "a new, previously undescribed, state." At a conference in 1959, he and Mollaret gave it a name: *coma dépassé*, beyond a coma.

Modern medicine was now challenging the familiar boundaries of death, just as it has changed how we think about birth. The beginning of life was once out of our control, until stem cell biologists learned how to turn an ordinary skin cell into an embryo, one that might potentially become a human being or something new like a brain organoid. There was a time when death was likewise inevitable if Bichat's vital tripod lost one of its three legs. Now artificial ventilation undermined Bichat's law, giving rise to a new kind of life.

Other doctors agreed with Mollaret and Goulon's worries about *coma dépassé*. "The developments in resuscitative and supportive therapy have led to many desperate efforts to save the dying patient," the Harvard anesthesiologist Henry Beecher said in 1967. "Sometimes all that is rescued is a decerebrated individual. These individuals are increasing in numbers over the land and there are a number of problems which should be faced up to."

There was a grim irony in the timing of *coma dépassé*. As ventilators were trapping patients in futility, transplant surgeons were learning how to save lives by moving organs from donors to recipients. In 1954 the Boston surgeon Joseph Murray replaced a man's damaged kidney with one from his twin brother. Finding people willing to give up a kidney was hard, and a donor's organ might not even be the right match for a patient. When it came to the heart or the pancreas, people didn't have an extra to spare.

Murray and other doctors turned to cadavers for more organs, but this method had its own drawbacks. Transplant surgeons had to make the arrangements to remove an organ while a dying patient was still alive. Then they had to wait until the patient's heart stopped and a doctor made an official declaration of death before rushing into surgery. The more time that passed between dying and transplantation, the more the organs deteriorated, and the worse the prospects became for the patients who received them.

Meanwhile, the transplant surgeons could see more and more patients lying in *coma dépassé*, awaiting death with their organs intact. "Patients are being brought in dead to emergency wards and potentially useful kidneys are being discarded," Murray complained.

Some doctors quietly took matters into their own hands. They would prepare a patient to receive a transplanted organ, and then they would wheel in another patient in *coma dépassé*. The doctors turned off the ventilator and waited for the sure sign of death: the ceasing of the heart. Then they immediately removed an organ from the donor and moved it to the living patient. This procedure cut down on the time required to carry out the transplant, but even in that short period the organs could still deteriorate.

In Belgium, a surgeon named Guy Alexandre decided he would no longer wait so long. Preparing for a kidney transplant, Alexandre picked out a patient who had suffered catastrophic brain injuries and showed no sign of brain activity. Without turning off the ventilator, he removed a kidney, which he immediately transplanted into its new host. The donor soon died, while the transplanted kidney went to work right away. It kept on working for the next three months, until Alexandre's patient died of sepsis.

In 1966, Alexandre described what he had done at a surgical conference. The other doctors in the audience balked. "I feel that if a patient has a heart beat, he cannot be regarded as a cadaver," said the British surgeon Roy Calne.

The president of the meeting asked for a show of hands from surgeons who agreed with Alexandre's definitions of life and death and would follow his example. Only one hand shot up: Alexandre's.

In 1967, Beecher organized a committee at Harvard to figure out how to define this mysterious new state. Murray and other doctors joined him, as did a lawyer and a theologian. They immediately launched into fierce debates for months, but eventually they agreed on a report that they published in 1968 in the *Journal of the American Medical Association*. It offered a new standard for declaring a patient dead: the death of the brain.

The committee argued that medicine had to free itself from obsolete notions of life and death. A stopped heartbeat had once been a reliable way to rule someone dead, since it led to the lungs and brain failing, too. Now doctors had the means to keep a heart beating even when the brain was damaged beyond hope. "These improved activities can now restore 'life' as judged by the ancient standards of persistent respiration and continuing heart beat," the committee wrote.

Beecher and his committee spoke of "life," rather than life. Massive brain damage often left patients without the remotest possibility of recovering consciousness, the committee declared. If doctors determined that their patients had "brain death syndrome," as the committee called it, the time had come to declare their patients dead.

The committee recommended that doctors carry out a series of tests before making the declaration. A patient's EEG reading should be flat. The pupils should be fixed and dilated. The doctors should shut off the ventilator for a few minutes to make sure a patient could not breathe without it. Some members of the committee thought doctors should repeat these tests for three days in a row. But the transplant surgeons found this delay too long with desperate patients waiting for organs. They persuaded their colleagues to knock the recommendation down to just one day. A doctor could then declare a patient dead and shut down the ventilator. The committee advised doctors never to reverse that order. "Otherwise, the physicians would be turning off the respirator on a person who is, under the present strict, technical application of law, still alive," they warned.

The committee's report was loaded with helpful guidance, but it was sorely lacking in argument. Beecher and his colleagues simply asserted that patients with brain death syndrome should be declared dead, rather than making a case for it. They raised huge questions, which they left hanging. When the committee claimed people with brain death syndrome had no hope of regaining consciousness, for example, did they mean that consciousness was the essence of life?

These gaps went overlooked when the report came out. The *New York Times* put it on their front page with the headline "Harvard Panel Asks Definition of Death Be Based on Brain." Doctors in the United States and other countries swiftly fell in line. A decade later a Harvard surgeon named William Sweet looked back at the 1967 meeting and judged it an unquestionable success. "The inescapable logic of the concept that death of the brain is equivalent to death of the person has now achieved widespread acceptance," he wrote. That acceptance gradually became law. States began adopting what came to be known as the "whole-brain standard": people with "irreversible cessation of all functions of the entire brain, including the brain stem," as the law put it, are dead.

On December 9, 2013, a girl named Jahi McMath was admitted to Children's Hospital in Oakland, California, for a minor procedure to treat her snoring. A surgeon removed her tonsils and part of her palate, and a few hours later she was awake and enjoying a Popsicle. But an hour later Jahi was spitting up blood. Less than five hours after that, her heart stopped.

Jahi's medical team rushed to her aid and got her heart beating again, but they had to put her on a ventilator. The following morning, when doctors examined her, they determined she had suffered a devastating lack of oxygen. No brain waves appeared on her EEG. Her pupils didn't react to light. Forty-five years had passed since Beecher's committee had mapped out the concept of brain death, and Jahi's doctors now decided she clearly met its standards. Three days after her disastrous surgery, Jahi McMath was declared dead.

Thanks to the ventilator, her lungs still inflated with air and her heart still beat. A social worker met with Jahi's stunned and devastated family to talk about shutting it off. But their experience with the medical staff had left them embittered. When Jahi started spitting up blood, her family had begged for help, and it only came slowly. Later, it emerged that the doctor in charge had made a note that her carotid artery was unusually close to her tonsils, which the hospital staff apparently overlooked. Now the social worker left Jahi's family feeling as if the hospital was leaning on them to kill her. They refused to agree to turning off the ventilator and instead asked for it to be kept running. And they also asked for a feeding tube so that Jahi would not starve to death.

The hospital refused to give care to someone who had just been declared dead. So the family went to court with their demand. "Plaintiffs are Christians with firm religious beliefs that as long as the heart is beating, Jahi is alive," their lawyer, Christopher Dolan, told the judge.

The judge ordered that an independent neurologist review the case. He came to the same conclusion as the hospital's doctors: Jahi was dead. Her beating heart was irrelevant; all that mattered was the state of her brain. After more negotiations, the family and the hospital reached an agreement. A coroner would issue a death certificate, and then the hospital would release Jahi to her family still hooked to a ventilator.

Using funds raised online, Jahi's mother, Nailah Winkfield, put her on a plane and flew with her across the continent. They landed in New Jersey, a state that allows families to reject brain death on religious grounds.

Most doctors and bioethicists found this turn in Jahi's case deeply frustrating. Brain death *was* death. Some experts hinted that moving Jahi's body to New Jersey was just a ploy cooked up by Dolan in order to squeeze money out of the hospital in a lawsuit. "She is going to start to decompose," the bioethicist Arthur Caplan told *USA Today*.

By the time Jahi was settled in at her new hospital, she hadn't eaten for three weeks. When doctors gave her a feeding tube, she began to improve. Most patients declared brain-dead died within hours or days, but Jahi remained alive week after week, month after month. Her teenage body was growing. She began to menstruate.

In August 2014, Nailah moved Jahi out of the hospital and into an apartment. Nurses cared for Jahi round the clock, and Nailah helped them turn Jahi every four hours to protect her from bedsores.

Meanwhile, Jahi's family launched a lawsuit against Children's Hospital Oakland for malpractice. A nonprofit foundation paid for a doctor to give her a new battery of neurological tests. Dolan later announced that the test revealed that some regions of her brain were still intact, with blood flowing through them. He asked a California court to have Jahi declared alive. Once again they turned his request down.

Three years later a staff writer for the *New Yorker* named Rachel Aviv paid a visit to the apartment. Nailah showed Aviv videos she had taken on her cell phone. In the jumpy movies, Jahi moved her fingers or toes, seemingly in response to her family and her nurses. When Nailah asked Jahi to move a finger, Aviv saw it move with what looked like a flicker of agency.

"I could also be investing undue meaning in gestures nearly too subtle to discern," she later wrote.

The case of Jahi McMath opened up a debate about the meaning of brain death. The more it unfolded, the more it became clear that the debate ran in striking parallel to the one over abortion. It came down to what we think it means to be alive—and, more particularly, what it means for us humans to live.

Ever since the Harvard meeting in 1967, some critics had questioned the logic of brain death, and Jahi's case now threw their questions in high relief. How could someone diagnosed with brain death have a heart that continued beating for years? How could she enter puberty, and possibly even respond to commands? Alan Shewmon, a California neurologist and a longtime critic of the brain death diagnosis, was invited by the McMath family to look over their videos and tests. "I am convinced that, from early 2014, Jahi McMath was in a 'minimally conscious state,'" he later declared.

Shewmon speculated that when Jahi had stopped breathing, her brain stem had been badly damaged, but parts of her cerebral cortex remained intact. That would mean she did not meet the whole-brain standard for brain death, even though her exams indicated she did. Shewmon speculated that the physicians who examined Jahi missed the fleeting moments when she could respond to the outside world.

Robert Truog, a Harvard pediatric intensive care physician, favored another possibility. Jahi really had met the criteria for brain death after her surgery in 2014. But she no longer did, Truog suggested.

"Perhaps McMath actually improved somewhat, rising a little on the spectrum of brain injury," Truog wrote in 2018. "This would not seem to be surprising in itself. But what makes this conceptually important would be that, in so doing, she would have crossed the bright legal line we have drawn between the living and the dead."

Other doctors were more skeptical. They were not impressed by the secondhand accounts of the cell phone videos. Still, no one denied that Jahi had gone through puberty. That transition is governed by the hypothalamus in the brain. Among its many tasks, the hypothalamus is responsible for releasing the hormones that trigger a child's body to mature. The fact that Jahi experienced puberty meant that at least this one small part of her brain was still intact.

The hypothalamus may be more resilient than the rest of the brain, thanks to its peculiar anatomy. It sits at the base of the brain, where it is nourished by a dedicated set of arteries. No one knows for sure how many other people who have been diagnosed as brain-dead had an intact hypothalamus. But there are hints that many did.

Among its other jobs, the hypothalamus manages the body's salt balance. It does so by squirting a hormone called vasopressin into the bloodstream. The hormone is exquisitely fragile, surviving only minutes after its release. To keep a body's level of salt steady, the hypothalamus has to monitor it and provide a steady supply of vasopressin. If a stroke or a tumor destroys the hypothalamus, it throws off the body's salt balance, leading to a condition called diabetes insipidus that can damage the kidneys.

In 2016 a team of researchers reviewed medical information on 1,800 patients who were diagnosed as brain-dead. Some of them suffered from diabetes insipidus, suggesting that their hypothalamuses were no longer working. But some did not have the condition. The researchers concluded that roughly half of the patients showed signs that their hypothalamus was still regulating their salt.

One of the authors of that study was Michael Nair-Collins, a bioethicist at Florida State University. He went on to publish a series of attacks on the whole-brain standard for brain death. He argued that a patient simply can't be in whole-brain failure if part of their brain—in this case the hypothalamus—is still working. If doctors come to that conclusion after an exam, Nair-Collins said, the problem does not lie in the patient's brain but in the exam, or perhaps with the concepts of life and death that the doctors rely on.

The hypothalamus is one of many parts of the body that are essential to keeping our body balanced. A proper balance of salt is important, but so is a proper blood pressure, which is regulated by hormones released by the kidneys. The body also needs a stable supply of red blood cells. The spleen destroys old cells as the bone marrow produces new ones. The immune system needs to fight off pathogens while brokering a peaceful relationship with the trillions of bacteria that live in our bodies. The food that comes into the body—whether through our mouths or through a feeding tube—has to be converted to sugar and other nutrients. The liver and other organs have to store extra sugar and then release it to keep its level steady in the blood.

In fact, the only reason that ventilators can work at all is that they pump air into bodies that are actively maintaining their inner balance. The air that they pump into the lungs has to reach the delicate ends of their branches, where their oxygen can get absorbed into blood vessels. The cells at those ends keep the lungs open by making a greasy film to coat the branched endings of the airway.

"The ventilator is capable of blowing air in and out of the bronchial tree," said Nair-Collins. "The organism must do the rest."

These facts apply not only to Jahi McMath, he argued, but to every patient diagnosed with brain death who can still breathe on a ventilator. In every case, the patient's body remains alive in a fundamental sense. "The implications for brain death are obvious," Nair-Collins said. "The patient meeting brain death criteria, supported with mechanical ventilation, is clearly biologically living."

While Nair-Collins called for abandoning brain death, its champions continued to fight for it. James Bernat, a neurologist at Dartmouth Medical School, published his first defense of brain death in 1981. When Jahi McMath's case drew national attention thirty-three years later, Bernat did not see it as a reason to give up the concept. The trouble lay not in the concept but in the tests. Jahi's diagnosis, Bernat said in 2019, "may represent a false-positive determination of brain death."

But one false positive didn't mean that the whole concept of brain death was wrong, Bernat argued. The cells in our bodies are alive, but our *human* life is not defined by its parts alone. What matters to a human life is how its parts are integrated, creating new levels of complexity. The human brain integrates signals from across the body and sends out commands to manage it. From that integration emerge our reasoning, our self-awareness, and all the other things that we call the human mind.

"Death is a biological and irreversible event that all organisms share in common," says Bernat. For all of them, death is the loss of their wholeness. A microbe dies if it loses the integration inside a single cell. For humans, Bernat argued, there's much more to lose: "Although both a living bacterial cell and a human being eventually die, the events of death markedly differ." In humans, the essential functions of the organism as a whole are carried out by the brain. So brain death—the permanent loss of those functions—is the criterion of death for our species.

As these arguments unfolded, Jahi began to falter. After three years of health, her liver failed and she developed internal bleeding. Exploratory surgery did not discover the source of her trouble, and she continued to decline. Jahi's doctors suggested another operation, but Nailah Winkfield decided Jahi had suffered enough.

"I told her, 'I just want you to know, don't stay here for me. If you want to go, you can go,'" she later said.

Jahi died on June 22, 2018, at age seventeen. Her mother brought her home at last to California for a funeral. The state of New Jersey, which had considered her alive until then, issued another death certificate. In the purblind eyes of the law, Jahi McMath died twice.

PART TWO THE HALLMARKS

DINNER

One afternoon in Tuscaloosa, I met a python named Haydee. At the age of three, she had already grown over six feet long, her muscular cross-section thicker than a bodybuilder's biceps. She lay coiled in a fiberglass box, and under its lights her scales gleamed like a dark diamond sleeve.

As I admired Haydee, her owner, a man named David Nelson, tossed her a live rat. The rodent froze in a corner of the box, but at first Haydee seemed indifferent to it. She gazed toward Nelson instead. It had been two weeks since she had eaten her last meal, so perhaps she wanted to see just how many rats were on the menu today.

Nelson moved on to tend his other snakes, and after a while Haydee lazily turned back to her visitor. She flicked her branched tongue. And then, in the middle of one of my blinks, she lunged. Her languid body turned into a missile.

On the roof of Haydee's mouth were a pair of long curved teeth. As her head crashed into the rat's body, she dug them into her prey. She coiled her trunk, wrapping herself twice around the rat. Above the coils, I could see pink legs and a hairless tail sticking up in the air. Between them I could see the rat's white midriff still taking in breaths.

It looked as if Haydee was suffocating the rat, but scientists suspect that's not actually how pythons kill. Their prey simply die too fast. It's possible that the snakes extinguish the life of their victims by pushing extra blood into their brains. Instead of a blackout, they experience a red-out. Haydee's rat grew still in under a minute.

She unspooled herself and slithered away as if she had forgotten the dead animal in her midst. Later she sidled languidly back. When she came face-to-face with the dead rat, she opened her mouth again. Now she used the small teeth on the sides of her mouth to grab its head. She did not swallow the rat so much as ratchet her own head over it. Saliva oozed from glands in her mouth to lubricate the rat's body, making it easier for her to slip her jaws over its shoulders and front legs. Her jaws stretched apart to either side so she could widen the passageway for her meal. Haydee pushed the rat down her esophagus by curving her body from side to side. After a few minutes of these contortions, she arched her head up, looking toward the glass door of the box again. She offered her human audience a chance to say farewell to the rat as its hind legs and tail glided out of view.

Barring a case of Cotard's syndrome, each of us knows that we are alive. Thanks to our socially tuned brains, we have a swift intuition about the lives of our fellow humans. It's harder to recognize life in other species, because we cannot talk to them or interpret a smile flashing over their faces. But from infancy onward we use mental shortcuts to sense the lives of others, like recognizing movements generated from within. At a young age children recognize that animals, like humans, are alive, but it takes longer for them to learn that plants are alive, too. As children get older, they don't lose these intuitions, but they do develop the ability to shroud them in words. If they're asked why they know a snake or a fern is alive, they will point to one of life's hallmarks—the things that seem to be shared in common by all living things. Children, in other words, are underage biologists. And biologists, in turn, are overgrown children.

I met Haydee on a series of trips I took to meet with overgrown children who explore life's hallmarks. Depending on the biologist you ask, you'll get a different set of hallmarks. But a few come up again and again: metabolism, information gathering, homeostasis, reproduction, and evolution. From species to species, each hallmark may take on an unimaginable diversity of forms. But underneath even the most extreme variations there's a unity.

I would not be able to swallow a rat whole like Haydee, for example, but I do need to eat in order to live. Hummingbirds need to drink nectar, and giraffes must browse treetops. A sequoia does not eat other living things, but it still eats after a fashion, its meals little more than air and sunshine.

This food then undergoes a transformation into work and flesh. Haydee turned much of her rodent meals into muscle, gut, brain, and bone. Sequoias turn their own meals into wood and bark. This transformation is known as metabolism, from the Greek word *metabolē*, meaning change.

I was introduced to Haydee by the person who understood the metabolism of pythons better than anyone: a biologist at the University of Alabama named Stephen Secor. I found Secor at his lab, and we took a drive from campus to the east side of town, past the Zion Hope Baptist Church and Moon Winx Lodge, until we reached the home of David Nelson and his wife, Amber. Secor pulled his RAV4 into the Nelsons' driveway just as David was heading toward his converted basement, hauling a blue cooler. He was a towering, bald man with a network of tattooed stripes that snooped out from the sleeves of his green T-shirt. The cooler was full of dead rats.

Secor and I followed Nelson inside. The basement's concrete floor was covered by black spongy squares. Half the space was given over to weight-lifting equipment and signs on the wall like *U.S.M.C.* and *TONY STEWART FANS ONLY.* The other half of the basement was full of stacked fiberglass boxes that looked like refrigerators tipped on their sides. Each box was fronted by a glass door, through which I could see a massive snake.

Nelson and Secor began taking snakes out of their boxes, letting them glide over their arms and necks. "How's my sweetheart?" Secor asked a python named Monty. "Monty's a good snake, aren't you?" he said.

"Oh, yeah," Nelson said gently, as if he were talking about his toy Pomeranian upstairs. Yet Nelson never let his guard down around his snakes. He was always aware of their movements, even as he let them flick their tongues across his eyebrow. "Any of these could kill you if you let it," Nelson said, somehow cheerfully.

Secor stood a few inches shorter than Nelson, but he could still handle powerful animals. He grew up working on a horse farm, thinking he would become a veterinarian. In college, one of his jobs was to help horses recover from surgery.

Horses have a bad habit of leaping back up before their anesthesia has fully worn off, only to stumble and break a leg. Secor had to keep them from rising before they were ready. He'd wrap his legs around their necks and use his arms to pin down their heads. At first the horses would be too groggy to resist him. Eventually they became strong enough to hurl him away.

"When they could throw me off, they had enough strength to stand up," Secor explained to me. It was during those horse-wrestling days that Secor decided to scrap plans of becoming a vet and go to graduate school instead, to study snakes.

Nelson was a product manager at a local car parts factory during working hours. The rest of the time, he was a snaker. He grew up catching snakes in the Alabama woods, and once he bought a house of his own, he started raising them indoors. He learned how to give a python a bath. He perfected a way to swiftly kill rats without suffering, by mixing vinegar and baking soda to flood a cooler with carbon dioxide. He learned how to groom a shedding snake so that its skin came off in a smooth sheet. He posted pictures of his pythons and boa constrictors on Instagram and brought them to his church's Bible school to teach children not to hate snakes. "At night, this is what I do," he said, looking across his serpentine kingdom.

Amber came down to the basement to watch the feeding. She had frosted blond hair and rhinestone hoop earrings. At first, she told me, she was not happy to be a snaker's wife. But she changed her tune when one of David's snakes got sick. Amber, who is a nurse, helped him keep the snake's nostrils clear so it could breathe. As the snake recuperated, it would coil contentedly in her lap as she watched television in the living room. "Momma mode came out, I guess," she said.

A friend in common introduced Nelson to Secor. At the time, Secor had some snakes that had gotten too big for his research, and he wanted to find them a good home. Nelson installed new stacks of boxes in his basement, and Amber translated Secor's soulless titles—AL1 and AQ6 and the like—into names like Haydee and Samson.

She was fond of all of Secor's snakes, save one. She dubbed him Lucifer. "You should have heard the first name," she said.

As Secor and Nelson let snakes glide around their necks, they rated the personality of each animal. Monty was good with children. Some of the other snakes were happiest in a dark corner. Others had figured out how to slide open the box doors and liked to crawl up to the ceiling fan. Delilah, an albino python, hadn't eaten for months. "She has these spells every year, and then she'll pound food," Nelson said.

Nelson got back to work feeding the snakes. Today the menu was rats, but on other days it was rabbits. "It's easier for time management," Nelson said. "I give them a rabbit and I'm done." Sometimes he managed to get weaned piglets for his biggest snakes. In the wild, pythons can readily eat prey half their weight. They have been documented eating deer and alligators.

Out of deer and alligators, pythons create fuel. It is the same fuel that we make, the same fuel that powers lichens growing on the tops of the Andes and the crabs skittering in the depths of the Pacific. It consists of molecules made up of carbon, hydrogen, oxygen, nitrogen, and phosphorus, known as ATP. Snakes and other animals make ATP inside their cells, using the sugar in their food and the oxygen they breathe. Plants use photosynthesis to make their sugar, which they can then use to build ATP. Some bacteria make ATP from sunlight as they bob on the sunny surface of the ocean. Deep underground, other kinds of bacteria make ATP by harnessing the energy stored in iron atoms.

Once living things build up a sufficient supply of ATP, they can use it as a fuel, breaking its bonds to unleash the energy inside. Haydee used it when she slithered around her box, powering the contraction of muscle fibers. She broke apart ATP molecules to power every beat of her heart. Her kidneys needed ATP to pull toxins out of her bloodstream. And the biggest entry in her fuel budget was the ATP she needed just to keep her cells intact.

Cells need to keep a big supply of charged potassium atoms on hand to carry out a lot of essential reactions. But with so much potassium inside a cell, there's a powerful force that pulls the atoms out into its surroundings. A cell would die if it just let its potassium bleed away. Instead, a cell uses molecular pumps studding its surface, each made from three interlocking proteins, to pull more potassium back inside. Just as a sump pump has to be plugged into a generator, a molecular pump needs ATP for power. It will use up one ATP molecule for every two potassium atoms it pulls in. These pumps have to run day and night, consuming vast amounts of ATP along the way.

Haydee's potassium pumps, like those in any organism, survive for only a few days before they start wearing out. Her cells have to rip them apart as they become defective and build new ones to take their place—a task that requires using up even more ATP.

The instructions for building new pumps are encoded in Haydee's DNA. Deoxyribonucleic acid, as this molecule is properly called, is made of two long strands that twist around each other. They're like a miniature spiral staircase with billions of steps. Pythons have 1.4 billion of these steps in their DNA, while we have over 3 billion. (Before you conclude that means we are genetically superior to pythons, bear in mind that onions have 16 billion.) Every step is built from two parts that stretch out from each strand. Those parts, known as bases, spell out instructions for molecules in a four-letter alphabet: adenine, cytosine, guanine, thymine; or A, C, G, and T for short.

Each of the three proteins in potassium pumps is encoded by its own stretch of DNA: a gene. To make a new potassium channel, a snake cell will bring enzymes and other molecules to the start of a potassium channel gene and read it one base at a time. They will produce a shortened, single-stranded readout, called messenger RNA. That molecule gets quickly sucked up by a floating cellular factory, which reads its bases and builds a corresponding protein. And at each stage of this creation, a cell must use more of its ATP.

Haydee was not just making new proteins to replace old ones. Her body was also growing. She had already tripled in size since she had hatched three years before, and she would keep growing throughout her life as long as she got to eat every few weeks. And she would need more fuel to expand. Just to create one new copy of its DNA, a cell has to break apart billions of ATP molecules.

Haydee even had to burn fuel to get fuel. She used up ATP to lunge at rats and choke them to death. She needed more ATP to build digestive enzymes, and those enzymes needed ATP of their own to break down the molecules in the rat. All living things face this same quandary: they pay a metabolic cost to keep their metabolism going. But snakes like Haydee take this quandary to an extreme. Pythons, boas, rattlesnakes, and a number of other species of snakes live through famines interrupted by feasts. They go for weeks without eating, swallow animals whole, and then extract as much ATP from them as possible in the days that follow.

Stephen Secor became fascinated by this living alchemy in the early 1990s. At the time, scientists knew very little about how snakes digest their prey. No one had even measured how much energy snakes used up in the process. Secor decided to find out, starting off with sidewinder rattlesnakes he caught in the Mojave Desert. He brought them to the University of California, Los Angeles, where he worked as a postdoctoral researcher. He fed them rats and then put them in a box.

The box was designed to measure a snake's metabolic rate: how much energy it used each hour. Secor took advantage of the fact that every time a snake uses up some ATP, it needs to make some more. And to make ATP, an animal needs oxygen. Every time a snake in Secor's box took in a breath, the level of oxygen around it dropped. From time to time Secor opened a stopcock on the side of the box, inserted a syringe, and drew off some air. The level of oxygen in the syringe told Secor how much ATP the snake inside was using up.

"In two days I had these numbers that made no sense," he told me.

After we eat a meal, our metabolic rate climbs by as much as 50 percent as we digest our food. The same is true for most other mammals. But Secor's rattlesnakes jumped by about sevenfold. With that observation Secor broke the standing record for the metabolic rate of a digesting animal. He then promptly broke that record when he switched his rattlesnakes for pythons. If he gave a python a quarter of its body weight in rats, its metabolic rate rose tenfold. Secor kept feeding some of his pythons until they ate their entire body weights in rats. Their metabolic rate increased forty-five-fold. For comparison, when a horse goes from a standstill to a full gallop, its metabolism increases by a factor of about thirty-five. But a horse cannot run at a full gallop for very long before getting exhausted. When a python digests a meal, it can burn fuel like a racehorse for as long as two weeks.

Now Secor was left with an even bigger mystery: How were these snakes ramping up their metabolism so dramatically, and what exactly were they doing with all that energy? The answer begins in the stomach, which makes hydrochloric acid to start breaking down food. Our stomachs release several squirts of hydrochloric acid a day, because we are adapted to regular meals. But a fasting python makes none at all. The fluid in its stomach is neutral, like water. As soon as Haydee swallowed her first rat on the day I visited her, her stomach received a signal to make a flood of fresh acid. By the time the rat's head reached the end of her esophagus, her stomach was ready to start dissolving it.

This acid flood was only one of many changes that Haydee went through once she caught the rat. Organs throughout her body began growing in order to handle the sudden onslaught of food. Secor discovered that a python's small intestines double in mass overnight, and the fingerlike projections on its cells extend their length sixfold. Once the partly digested rat arrives in the intestines, they are ready to absorb glucose, amino acids, and other nutrients and deliver them into the bloodstream. The liver and kidneys also double in weight in advance of the work they'll have to do to store the nutrients and expel the waste. The heart grows 40 percent in order to push the extra load of sugar and other nutrients around the body.

But this discovery only left Secor more baffled. He had no good answer for how the snakes were transforming their bodies. They have the same basic anatomy and biochemistry as other vertebrate animals. They have livers, stomachs, and hearts that work much like ours do. Their cells come in many of the same types as in our bodies, from neurons to pathogen-killing immune cells. Many of their genes are nearly identical to our own, encoding the same hormones, neurotransmitters, and enzymes. Secor suspected that snakes are able to transform themselves so drastically not because they have unusual genes, but because they use their genes in an unusual way. Their genetic orchestra uses the same instruments as ours, but they were reading different sheet music.

When cells have to carry out a job—whether fighting a virus or secreting bone—they begin reading certain genes and making proteins based on their sequences. Some of these genes encode proteins that act like master switches: they latch onto other genes and switch them on. Some of those genes encode even more master switches. A regulating protein may ultimately trigger hundreds of genes, which produce a swarm of proteins that together carry out complex jobs. Secor formed a hypothesis that snakes are using their regulating proteins in a peculiar way. But in order to test that hypothesis, he would have to track the activity of genes in his snakes. In the early 2000s, when Secor started looking around for help, geneticists told him he was on a fool's errand.

"I said, 'What would it take to go after this?'" Secor recalled. "And they'd say, 'You couldn't do it. It would take years and years and years, because you'd have to pull each one out, and then you'd have to find out what it was.'"

In 2010, Secor finally found someone who didn't shut him down: a geneticist named Todd Castoe. At the time, Castoe was sequencing small chunks of DNA from reptiles at the University of Colorado School of Medicine. Secor and Castoe became scientific partners, bringing together a team of researchers to sequence the entire genome of the Burmese python. Once they had completed the job, they had both a catalog and a map to guide their research. Now they could track the activity of genes as pythons changed their bodies for digesting.

Castoe and Secor began collecting muscle and other tissues from pythons, fishing out the messenger RNA the cells were making. They looked at their catalog to match up the messenger RNA to the genes in the python genome. Secor and his students compared the activity of genes in snakes before and after eating to look for changes brought about by their metabolism. The researchers expected that maybe twenty or thirty genes would switch on. Instead, the snakes performed a far greater transformation.

Within twelve hours of swallowing a rat, the scientists found, the pythons switched on several thousand genes in organs across their bodies. Many of the genes worked together along ancient pathways, the same pathways found in many other species of animals. Some of the pathways that become active when a snake eats are used by many animals to make their bodies grow. Others let them respond to stress. And still others create the proteins necessary to fix damaged DNA.

The growth pathway genes may enable snakes to swell their organs to prepare to metabolize a massive meal. But making billions of new cells in a matter of hours may also harm the snakes. Their cells grow so quickly that they make deformed proteins, which create stress. Charged molecules may fly around the cells, damaging their DNA. The snakes have to fix this damage, a task that increases the metabolic cost of digestion even more.

Haydee would spend the next week or two digesting today's meal. All told, she would burn up about a third of the energy in her rat simply to digest it. She would burn so much fuel so fast that her body temperature would rise. On an infrared nightscope, she would look as warm-blooded as a live rat. It's not a waste, though, since her metabolic fires still leave two-thirds of the rat for herself. Her blood would course with fatty acids, at a concentration high enough to kill a human. Across her body, her cells would take up the calcium, the amino acids, and the sugar she harvested from her prey. She would grow some more muscle, add bone to her skeleton, and store away new fat.

And to survive until David Nelson gave her another rat, Haydee would undo all the fleshy equipment she had swiftly created to digest her meal. Her strange, borrowed genetic networks would shut down. Her organs would shrink down to their previous size. The cells in her intestines would retract their tentacles. She would excrete all that was left of the rat, a pod of hair, as she entered another long fast. There's a simple logic to this extreme cycle, but it is so far from our own experiences that it can seem bizarre even to trained scientists. Secor would sometimes show pictures of the intestines of fasting snakes to pathologists. He'd point out the shrinking fingers and ask their opinion of what was happening to his snakes.

"Your animals are sick. They're dying. They have parasites that are ravaging their intestines," they'd say.

"No, they're healthy," Secor would insist.

Secor never managed to persuade the pathologists that they were just looking at a different kind of metabolism. "They just shook their heads and sent me on my way," Secor said.

By the time Secor and I were ready to leave the Nelsons' house, Haydee had become still, a loosely coiled rope of glistening biomass. I could hardly make out the bulges where the rats she had swallowed were gliding toward digestive destruction. It was hard to believe she was the metabolic equivalent of a racehorse. In a few days, when she had finished absorbing their nutrients, her metabolic rate would slide down again. She would still need to burn a little fuel to beat her heart, to pump charged atoms in and out of her cells, to grow a little longer. Her basal metabolic rate would never drop to zero. But she would get remarkably close.

DECISIVE MATTER

Subash Ray pulled open a drawer and took out a smudged piece of paper. It looked as if he had spilled his coffee on a Post-it note, ignored it for a few days, and then tossed it in the drawer instead of the trash. But now Ray was going to perform some magic on it.

"We are going to bring spring to its life," he said.

Ray had a round face and rectangular glasses. He wore jeans and a polo shirt decorated with a tiny dark eagle. He spoke softly, so softly that sometimes I had to ask him to repeat himself as he described to me what he was doing. I came to visit Ray and his colleagues in the city of Newark, where he was earning his PhD at the New Jersey Institute of Technology by studying these smudges and what they can become.

Ray stretched his arm up to a high shelf and grabbed a jar of dried algae extract, known as agar. He put it on the seat of a lab chair as if he were in a supermarket and it was his shopping cart. He added the smudged paper, which he stored in a jewel glass for safekeeping. He found a pair of beakers and a kitchen whisk.

His chair fully loaded, Ray wheeled it into another room in the lab. I trailed behind, along with Ray's supervisor, a biologist named Simon Garnier. Garnier was a red-bearded Frenchman who wore a hoodie and played European handball—"like water polo on land," he said, in an unsuccessful attempt to explain it to a confused American.

Ray rolled up to a sink, where he filled up an electric teakettle and switched it on. He put his beaker on the counter and filled it with the water after it had gotten hot. His whisk clanged as he stirred in some agar, and then he poured the mixture into an empty petri dish.

Once the agar cooled to a firm bed of goo, Ray picked up a pair of tweezers, plucked the smudged paper out of its jewel glass, and transferred it to the petri dish. He tamped it down on the agar and gave it a spritz of water.

Now Ray steered the chair away from the sink, to a windowless room that was overheated and sticky with humidity—the conditions in which many living things like to grow. Against each wall there were large white boxes sitting on tables. Ray turned a knob on the front of one box and swung it up like a door. Inside I could see a pair of metal rails on which rested three downward-pointed cameras, complete with flashes. Ray slid the dish, with its smudge and goo, under one of the cameras.

Garnier sat down at a laptop and began typing commands. After a few moments the box lit up with a white glow, and then the camera flashed. As the box turned dark again, we left the room, the camera set to snap a new picture of the dish every five minutes.

That evening Garnier took me out to dinner with some of his fellow biologists. We walked down Raymond Boulevard, a street full of human life and human construction—of little nail salons and massive warehouses, of empty Art Deco buildings for lease and bus stops crowded with waiting passengers. We reached a boutique restaurant where we sat around a wooden table and shouted over the din about the living things around which each of the biologists had built their working lives. They talked about the nervous systems in worms the size of a comma, about the transparent bodies of zebra fish. Meanwhile the camera back at Garnier's lab popped flashes through the night.

The next morning I made my way back to the lab in the Central King Building, and we reentered the humid camera room. We looked at the dish. The smudges were gone, and in their place was a lemon-colored blob. The blob had spilled over the edges of the paper and spread across the dish, growing to the size of a silver dollar. Garnier smiled as he looked over the change.

"Well, they're alive," he said. "They haven't moved much, but they're alive."

When I inspected the blob, I could see that it was actually a thicket of tentacles that branched again and again as they radiated out from the center of the dish. They were still splitting even as I looked down at them, but too slowly for my short-attention brain to perceive.

The creature Ray had brought to life was *Physarum polycephalum*, otherwise known as the multiheaded slime mold. You can't find slime molds on the streets of Newark, but travel a few miles to a wooded preserve beyond the city—to the Eagle Rock Reservation or the Great Swamp—and on a suitably warm, damp summer day you can spot its golden web on decaying logs or the cap of a mushroom. Pretty much anywhere on Earth where a forest grows, you can find *Physarum* or one of the hundreds of other species of slime molds. Their bizarre appearance inspires visceral names. Wolf's milk. Dog vomit.

After growing over the summer, *Physarum* prepares for winter by making spores. The spores survive the cold as the rest of the slime mold turns to a dead black crust, and in the spring they start growing again. But if this cycle is interrupted by a catastrophe—

a drought or the crash of a tree exposing the forest floor to harsh sunlight—*Physarum* takes emergency measures. Its entire body dries up into a drab, brittle form, called a sclerotium. The sclerotium flakes into fragments and blows away. If one of those fragments lands on a damp patch of ground, it revives. Slime mold researchers can make a sclerotium simply by putting a dollop of live *Physarum* on a piece of filter paper and letting it dry out. They can store it away for weeks or months. If they then put a sclerotium in a dish of agar, they can bring spring to its life.

Flashing through the night, the camera had created a stop-action movie of Ray's slime mold. Sped up for human perception, it revealed the smudge turning to gold before ballooning off the edge of the paper and expanding across the agar. Later in the night, the tentacles on the opposite side of the paper spread out as well. Now the slime mold became a spreading disk.

Its movement was not the result of gravity acting on passive matter. The slime mold did not spread like a drop of water. It was displaying a hallmark of life: it was using its own stores of fuel, its own proteins, the logic encoded in its own genes—the same combination found in all living things—to make decisions about what to do next. It had agency. It was hunting.

The graduate students and postdoctoral researchers who worked with Garnier were an eclectic band. Some were off in Namibia, putting collars on baboons to track their movements and record their grunts. They were learning how the baboons stayed together as a group by exchanging information about their locations. In Panama another student was studying how millions of army ants used their bodies to create a living nest, including chambers where their queen could live. The concept of decision-making conjures up a human brain—plump, convoluted, generating word-laced thoughts about the future. Our brains are tens of thousands of times bigger than the brains of ants, and yet ants can together create a house out of their own bodies. Slime molds, without any brain at all, distill life's decision-making down to an even more exquisite essence. "What I really like is that it gets back to the origin of intelligence," Garnier told me.

In the forest, a slime mold searches for bacteria and spores of fungi. It extends its tentacles across logs and soil until it finds prey. As it crawls over its victims, it oozes out cell-slashing enzymes and drinks up the debris. "It's a moving stomach," Garnier said.

When Ray brought his slime mold back to life for me, it started looking for food, but he had not provided it any food to find. To let me see how a slime mold discovers its next meal, Ray set up a new experiment. He placed three pale chunks of cooked oatmeal on the agar, like the corners of a triangle.

"If you can cook porridge, you can grow slime molds," Ray said. I glanced up and noticed the rows of Quaker Oats canisters on the lab shelves. They looked down at the scientists with cheerful colonial smiles.

"They like the old-fashioned ones," Garnier said—*they* being the slime molds. To be more precise, they like the bacteria that grow on the old-fashioned ones. No breakfast is sterile.

Ray dropped a dollop of live *Physarum* in the center of the dish. It could not see the pucks of oatmeal. But it could taste the sugars and other molecules that diffused away from the food and spread through the agar. As the slime mold's tentacles spread out from the center, proteins on their surfaces picked up these signals. It then used a simple set of rules to seek the food.

As each tentacle moved, it compared the concentration of molecules at different points in its path. If the concentration dropped, the slime mold stopped extending tentacles in that direction. If the concentration went up, it continued to explore. Hours after

Ray dropped the slime mold in the middle of the dish, its tentacles had reached all three pucks. As they infiltrated the oatmeal, they turned it from gray to gold.

With no brain to issue commands, slime molds allow scientists to see how life's powers of decision-making can emerge from nothing more than biochemistry. They've discovered an entire playbook of elegant rules that slime molds use to thrive. To show me one of their more impressive tricks, Ray re-created an experiment that was first carried out in 2012 by another former student of Garnier's. He created a slime mold cul-de-sac.

It was simple enough to build. Ray used scissors to cut a sheet of acetate into a sharp-cornered shape: ⊔. He set the acetate in a dish. Slime molds can only crawl over damp surfaces, which meant that the dry acetate was as impassable as a high brick wall.

Ray then placed a scoop of *Physarum* near the open end of the cul-de-sac. On the opposite side of the dish, he placed a drop of sugar. The wall of the acetate cul-de-sac lay between them, but the sugar could sneak under it, diffusing through the agar, teasing the slime mold with its fragrance, luring it into the trap.

The next day, when we came back to look in on the slime mold, it had escaped its cul-de-sac. When I watched the overnight movie, I felt like a guard reviewing a prison break. The slime mold followed the trail of sugar into the cul-de-sac and hit the acetate wall. But it did not give up its search. It sprouted tentacles to either side. Its left-hand branches eventually reached the corner of the wall and then turned back, exiting the trap. They then wheeled around and made their way along the outside of the wall, heading toward the sugar.

Slime molds manage this escape by using a brainless kind of memory. They continually send out probing tentacles, and the ones that don't detect increasing signals of food retract. As they pull back, the tentacles leave behind a slimy coat. *Physarum* can sense its own trails and will steer its new tentacles away from them. This external memory lets a slime mold override its attraction to sugar. Rather than banging its multiheaded head against the acetate wall, it can move out of the cul-de-sac and explore new paths toward food. We need a brain to remember things, but *Physarum* has no such organ. Instead, it stores a record of its experiences in the outside world.

Slime molds have solved far more complex problems. A scientist named Toshiyuki Nakagaki, for example, found that slime molds can discover the shortest path through a maze. He built a labyrinth by cutting pathways out of a plastic sheet and setting it on a bed of agar. Nakagaki and his colleagues placed an oat flake covered in slime mold at one opening of the maze and put more oats at the other. The slime mold extended new tentacles through the labyrinth, exploring every possible path. Once it found the oatmeal at the end of the maze, it began feeding on both supplies of food at once as it retracted the branches in the dead ends. Eventually the slime mold became a single streamlined tentacle mapping a route through the maze. Nakagaki designed the maze so that the slime mold could take four possible routes to the food. In the end, he found, it always traced the shortest path.

Some scientists have given slime molds other puzzles to solve that have more bearing on their lives on the forest floor. In nature, slime molds don't find food at the two ends of a maze. Instead, they may encounter patches of food scattered across a log. If they can feed on all the food at once, they will grow faster. But, to reach all the food, they have to pay the metabolic cost of building tentacles. If they overbuild, they will use up more energy than they get from the food.

It turns out that slime molds are very good at finding an efficient solution to this problem: they work out the shortest path to several pieces of food at once. Nakagaki and other slime mold experts have run experiments to see how slime molds make these complex choices. They sprinkled oats across a dish and watched *Physarum* work out a solution. Instead of forming a single zigzag tube, it built a network that connected the oats in close to the shortest possible distance. In one experiment, scientists created a map of the United States, with oatmeal standing in for the biggest cities. The slime molds built what looked remarkably like the American interstate highway system. They have mimicked Tokyo subways and Canada's transportation network. It's unsettling to mathematicians that slime molds can solve this kind of problem in a few days. It's kept them busy for centuries.

Another puzzle that has kept generations of mathematicians busy is known as the Knapsack Problem. Imagine you're preparing for a hike, and you have to decide what to put in your knapsack. You can choose from a lot of different items that are more or less useful for the trip. But you also have to keep in mind the weight of your items, since you can't pack an infinite number of them. You might tuck a deck of cards in your backpack so that you'd be able to pass the time on rainy mornings in the mountains playing poker. But you wouldn't fill your knapsack with a forty-pound chess set made of carved soapstone merely to make sure you didn't get bored. Mathematicians distill this choice into a pure abstract form. You have a set of items, each with a value and a weight. Now you must find the combination of items with the greatest value that is under a certain weight.

Many businesses face practical versions of the Knapsack Problem. Airline companies want to figure out how to pack their planes so they can deliver the most valuable cargo using the least amount of fuel. Financial firms look for the best way to spread their funds across investments with different potential returns. There's no simple equation to solve the Knapsack Problem, though. Researchers have filled books with strategies to get us close to the best solution.

Slime molds may not be able to write books, but they can solve the Knapsack Problem. Audrey Dussutour, a scientist at Paul Sabatier University in Toulouse, France, and her colleagues brought their skill to light by translating the problem into the terms that matter to slime molds: food. To grow as fast as possible, *Physarum* needs both protein and carbohydrates. It turns out the optimal mix is two parts protein to one part carbohydrates.

Dussutour offered *Physarum* a choice between two lumps of food, both far from ideal. One was nine parts protein to one part carbohydrates, and the other was one part protein to three parts carbohydrates. If the slime mold reached out to the first lump and only ate that, it wouldn't get enough carbohydrates. The second lump would leave it without enough protein.

The slime mold managed to turn Dussutour's two bad choices into one good one. It grew tentacles out to find them both. Eventually its network collapsed into a highway joining the two supplies of food. Simply mixing them together would not give the slime mold its ideal diet. And so the slime mold drew more food from the protein-rich supply than the carbohydrate-rich one, balancing out its meal close to its ideal two-to-one ratio. In other experiments, Dussutour tried out more combinations, and the slime molds always figured out how to balance them. In other words, they learned how to stuff their knapsack with the right blend of supplies.

As slime mold researchers run more of these experiments, they better understand how a web of *Physarum* makes its way in a forest. It takes in information about everything it touches, and as it encounters places rich in bacteria and spores, it can shift to these feasts. If it creeps out into the sunlight, it can pull back into the shade. It can adjust its network from day to day with a mathematician's precision so as to guzzle the biggest meal for the least cost. It's a strategy that works impressively well. Under the right conditions, slime molds can get as big as a throw rug.

When I asked Garnier how exactly slime molds solve all these problems, he gave a Gallic shrug. "Welcome to the beautiful world of slime molds, where nobody knows much," he said.

But one of his graduate students, Abid Haque, was willing to show me where he and Garnier suspect that some answers lie: inside its golden tentacles.

Before coming to Newark, Haque had been studying to be a mechanical engineer at the Indian Institute of Technology Guwahati. A summer research project lured him to the kingdom of *Physarum*, and now he was working toward a PhD in Garnier's lab. The day we met, he was wearing a black T-shirt covered with Victorian engravings of slime molds: filigreed spore cages, tadpole-like sexual cells, and *Physarum* webs looking like elastic trees.

Haque carefully snipped an inch-long slime mold tentacle and took it to a dim microscope room. He quietly twirled the knobs on the microscope for a few seconds. "Oh, this is gorgeous," he said.

When I looked down at the slide, it took a while for my eyes to adjust and for my brain to figure out what I should be seeing. And then, in an instant, I was looking at a green river. The current carried along grains, some dark and some light. As I watched, the river slowed down. The grains coasted to a stop. After a moment of stillness the river reversed course and pushed the grains back the other way.

The lighter grains contained enzymes that the slime molds use to break down food. The darker grains were nuclei, microscopic sacs in which they hold their genes. We have nuclei in our cells, too, but typically each cell contains just one. When it divides in two, it makes a new nucleus so that each new cell inherits its own set of DNA. Slime molds can also make new nuclei, but they don't bother to divide their cells in two. Instead, every slime mold—stretched out over a petri dish or across a forest floor—is a single gigantic cell.

"It's mind-boggling that it's just one thing," Garnier said.

Physarum comes from the Greek, meaning "small bellows." It was probably inspired by the pulsations that naturalists could see with their naked eye in the slime mold's golden web. The early generations of slime mold scientists had no way to determine what made them throb. It was only in the 1900s that biologists got their first glimpses of the molecules that slime molds are made of.

Each tentacle is enmeshed in a microscopic skeleton of wires. It's not a rigid trusswork like the Eiffel Tower, though. A slime mold is perpetually building new parts of its skeleton and taking down others. It can assemble wires in a tight network that pinches a tentacle and pushes away the fluid inside. If the wires slide away from each other, the walls of the tentacle relax, letting the fluid flow back in.

Squeezing and slacking, the slime mold beats like a web-shaped heart. The pulsations push the grains in waves, and these waves can ripple across its entire network, crashing into each other to create even more complex patterns.

Haque and Garnier wondered if these waves might serve as a kind of information relay for the single-celled slime mold, allowing it to learn about its surroundings and merge its findings in a vast, wave-based computation. The slime mold could then reach decisions about what to do next.

To decipher this language of waves, Haque began with a simple experiment. He placed inch-long sections of slime mold tentacles in dishes. Inside each tentacle, waves traveled back and forth. Haque put a dollop of food just out of reach of each end. One was rich with oatmeal, the other less so—and thus less desirable. The slime mold sensed the food and stretched out in both directions. And as it probed the two meals, Haque found, its waves changed.

Haque and his colleagues found that the waves moved more often toward the good food than toward the bad. And as the waves changed, the slime mold itself changed. The skeletal wires at the end that was feeding on the good food fell away, causing the slime mold to swell. Meanwhile, some researchers have theorized, the end of the tube feeding on the bad food stiffened its walls. The result was that the slime mold crawled away from the bad food and engulfed the good.

"It would be like when you get to a good location, your muscles melt," Garnier told me. "But it's okay, because you are in a good place."

Melting your muscles once you reach a good place is, Garnier argued, a kind of intelligence. To him, intelligence is not a score on an IQ test or the ability to learn Dutch. It is a hallmark of life: the ability to respond to a changing environment in a way that helps keep an organism alive.

"If you compare any organism, they do better than random," Garnier said. We need an overstuffed brain to do better than random, but cellular waves rolling across a network of tentacles may also suffice.

"The slime mold," Garnier said, "is the thing that has pushed this principle as far as possible."

PRESERVING CONSTANT THE CONDITIONS OF LIFE

On a snowy morning in the Adirondacks, I hiked up a hillside to an abandoned graphite mine. I trailed a pair of biologists, Carl Herzog and Katelyn Ritzko. They stopped at the maw of a mine, alongside a frigid stream.

Herzog and Ritzko began to change gear to go inside, and I followed their example as best I could. I took off my hiking boots one at a time and tried to get my socked feet into a pair of chest waders without tripping over into a drift. We put on helmets with headlamps. We peeled off outer layers of flannel and fleece. It was time to trade our defenses against the snow and winter winds for defenses against cold water and sharp rocks. Herzog ran down a list of risks we might encounter inside the mine. "Tripping and falling is the biggest threat," he said. "You don't ever want to touch the ceiling."

As Herzog spoke, Ritzko was storing pencils and notebooks in her wader pockets, checking batteries in an assortment of devices.

"Ready to go?" Herzog asked her.

"Tally-ho," she replied. We stepped down into the frigid current and waded into the mine.

The snowy light dimmed as we splashed forward. The walls were sloped and jagged. After a night of hard rain and a morning of snow, water was pouring down from the overhanging hillside. As it streamed into the mine, the water froze into icy versions of stalactites and stalagmites. The cave grew dim as we waded farther in, and the stream gained a thick frozen lid, as clear as a window.

Ritzko climbed out of the stream and walked along a narrow strand of loose rocks running along the right-hand wall. But Herzog wanted to take a closer look at the left side of the cave. He stepped up onto the transparent ice, levitating one nervous step at a time.

"If I break through, I have waders up to my chest, so it should be no concern at all," he said, hovering. "But it's amazing how much anxiety it conjures."

Herzog scanned a flashlight across the wall and saw nothing. He gingerly crossed back over the skin of ice to join us, and we moved into the darkness.

I had to remind myself I was not in a natural cave but a vast man-made hole. In the mid-1800s, lumberjacks around Lake George noticed veins of dark minerals in their log skids. They turned out to be deposits of graphite. The lumberjacks reinvented themselves as miners, digging into hillsides and hauling out graphite to be turned into pencils and crucibles. The mine that we were now entering, near the town of Hague on the banks of Lake George, grew over the years into a network of rough-walled tunnels, tight passageways, and side chambers. As the miners carted out rock, they sometimes brought in tall timbers to prop up the ceilings, creating an underground grove of undead trees.

The New York graphite boom lasted a few decades before bigger, cheaper mines opened up in Madagascar and other countries. In the early 1900s, miners pulled out some of the timbers from the Hague mine to sell for lumber and then gave up on it completely. Over a century had passed, and now I could see only a few traces of their presence. In the surrounding woods, there were scattered hillocks of rock hauled out of the mine. A string ran down the length of a deep tunnel, perhaps a guide back out for men lost in the dark.

After people abandoned the mine, the elements slowly reclaimed it. Water coated the hacked walls with a glossy coat of flowstones. It decorated the ceilings with striped ribbons known as cave bacon. Some of the timbers that the miners hadn't pulled out had since tumbled into the stream. Herzog pointed out parts of the ceilings and walls that had collapsed, dumping fresh rock and erasing passageways.

"I don't expect it to happen while we're here," Herzog said, but he warned me not to touch the timbers. "You can dislodge things if you touch them."

Herzog and Ritzko's flashlight beams flicked over the walls and ceilings, up into alcoves, deep inside fissures. This cold, rocky labyrinth seemed about as lifeless as a place could be. But after an hour of spelunking, Ritzko's beam froze. I made my way over the loose rock to where she stood and followed her gaze. At eye level I saw what looked like a furry pear dangling from sheer stone.

"It's a northern long-eared bat," Ritzko whispered.

The bat pressed its face against the cold mine wall. I could make out its wedge-shaped ears poking out from its head and its miniature feet splayed out as anchors.

"How does it stay up?" I whispered.

"Their ankles have a locking mechanism," Ritzko said. "It expends almost no energy to hang there."

"Is it even breathing?" I wondered.

"It *is* breathing," Ritzko whispered. "But everything is much slower."

The bat we observed had flown into this mine four or five months before. It had found a place to hang from the walls of stone and had survived through the winter without eating a morsel. Within a few weeks the bat would fly out again, to enjoy months of glorious spring and muggy summer. Even on the hottest days it would manage not to cook itself. It might get infected with bacteria and fend them off. It might have a bad night of hunting and manage to avoid starving to death. When it chased after prey, its heart could pound without pushing so much blood into its head to make its brain explode. And in the fall, it would return to another mine like this one to endure another winter. As Ritzko and I examined the northern long-eared bat hanging before us, I marveled at the fact that this animal can withstand these unpredictable crises and drastic crises for eighteen years or more. The bat and the mine in which it hibernated could not be more different. The lifeless mine was gradually collapsing. Within a few decades the eroding power of the seasons might erase it altogether. And yet this little bat inside the decaying mine remained astonishingly stable.

"All the vital mechanisms, however varied they may be, have only one object, that of preserving constant the conditions of life in the internal environment." So wrote the French biologist Claude Bernard in 1865. Bernard observed that our internal environment is mostly water. When our body's water supply starts running low, we get thirsty, causing us to replenish it. In 1926 the Harvard physiologist Walter B. Cannon updated Bernard's concept and gave it its modern name: homeostasis.

Homeostasis is not a physical thing to be weighed or poked. It is not a particular assembly of atoms forming a molecule like DNA or proteins. It's instead a principle that you can find throughout the living world, acting on many levels at once. In a bat, it is present in its cells, in its organs, and even in its flight.

While few people have seen a bat hibernating in a cave, many have seen them on warm evenings, darting through the fading light after a mayfly or a mosquito. Once the sky grows black, bats will go on flying, unseen in the dark, traveling hundreds of miles in a single night. They stay aloft thanks to an airborne form of homeostasis.

To fly, bats flap their gargantuan, membranous hands. As they push down, they set the surrounding air into swirls that circulate around their wings, and with each upstroke they shed some of the air behind them as spinning donuts of air. The physics of these swirls is so complicated that scientists still barely grasp the fundamentals. But the outcome is clear: the pressure above the wing drops as the pressure below increases, creating an upward force.

By adjusting the timing of its wingbeat, by spreading or closing its long, spar-shaped fingers, by contracting some wing muscles and relaxing others, by sculpting the invisible donuts of air that trail behind it, a bat can precisely cancel out gravity. It hovers in place. And if a hovering bat tilts its wings, it can turn some of that lift into thrust, which shoots it forward. Northern long-eared bats and other insect-eating species chase their prey by shrieking and listening to the echoes that bounce back. Many of the species they hunt have evolved the ability to hear bat echolocation, and they try to bank suddenly to escape. The bats can follow suit, folding a hand to make a tight turn.

As a bat flies, it perpetually runs the risk that the swirls of air will peel away from its body, causing it to fall like a pebble. A homeostasis of air keeps it aloft. One secret to this stability is a scattering of tiny hairs on the bat's otherwise hairless wings. The hairs sway as the currents of air shift around them, and their fluttering gets translated into electric signals that travel to the bat's brain. The bat can sense warning signals that swirls of air are going to peel away and adjust the shape and curve of its wings so they will keep hugging its body.

Bats regularly get buffeted by unexpected gusts of air. When swarms of them rush out of caves each evening, they often crash into each other. Because bats are so small—a northern long-eared bat weighs about as much as an empty envelope—these disruptions can easily throw them off kilter. They may then stall and crash.

The biologist Sharon Swartz got to wondering why bats aren't constantly dropping out of the sky. At her lab at Brown University, she and her students filmed bats in flight, capturing a hundred images every second. To study how they handled gusts, they installed a tube that could deliver a puff of air to bats flying past. It hit one of their wings, swinging their bodies about a quarter turn.

In less than a tenth of a second, Swartz found, the bats righted themselves. A close look at the films revealed their trick. If a puff of air caused a bat to roll to the left, it stretched out its right wing, forcing its body to roll back. As the bat approached an even keel, the rotational forces of the two wings perfectly canceled each other out. Its balance was restored.

This strategy is familiar to engineers. It's the same one built into cruise control systems, for example. When a driver puts a car on cruise control, the vehicle doesn't simply spin its engine a fixed number of times a second. It continually adjusts the rate as it senses its acceleration. If the car heads down a hill, its sensors cause it to slow down. Once the car drops below the desired speed, it gently accelerates again. Negative feedback loops, as engineers refer to these designs, keep systems stable by pulling them back to a set point each time they're disturbed.

Bats use negative feedback loops not only to stay in flight but to keep their chemistry in balance. The sugar in their bloodstream remains exquisitely stable even as they feast on insects, burn fuel in flight and then fast during sleep. When bats sense that their blood sugar is rising, a blast of insulin triggers their cells to store away the extra supply. If the level of sugar in the blood drops a little, the cells release enough to bring it back up without overshooting.

Bats have other negative feedback loops for their salt, potassium, and acidity. Like humans and other vertebrates, bats have a circulatory system powered by a beating heart that demands a steady pressure to work. To stay at that set point, bats use negative feedback to make their blood vessels relax and tighten. Bats also keep their bodies at a constant temperature. If they overheat, they can pump extra blood into their skin to dump excess heat into the air. Cold bats burn fat to stoke their metabolic fires.

Bats evolved perhaps 60 million years ago, on a planet so warm that Antarctica sprouted forests. Today most of the 1,300 species of living bats are limited to the tropics. But some, like bats we were observing in the Hague mine, have adapted to life closer to the poles, in places where they have to endure long winters without any insects to catch, nectar to sip, or fruits to nibble. Making matters worse, the cold temperatures in winter demand extra energy to keep their bodies warm.

Bats have evolved an extraordinary strategy to thrive in these unwelcoming places, one that other species such as black bears and ground squirrels have also hit on: they hibernate. In other words, the bats reset their homeostasis around a new set point.

After a busy summer of hunting, northern long-eared bats search for mates by visiting new caves or mines. Each evening they fly out into the darkness again to find more food to store away in their bodies to get through the winter. A six-gram northern long-eared bat may put on an extra two grams of fat. Imagine surviving a five-month famine on half a teaspoon of butter. The bats choose one last cave or mine where they will spend the winter—a hibernaculum. They clamp their feet to the walls, hang upside down, and slow their breathing. Within an hour their body temperature plunges. They become as chilly as the cave air around them.

As Ritzko and I inspected our northern long-eared bat, she jotted a few notes down on her pad. When she was done, she cast her flashlight farther down the mine and soon found more. I let her go, and clambered over to Herzog, who was finding bats of his own. The mine was home not just to northern long-eared bats, but to other species: little brown bats, big brown bats, small-footed bats, even the rare tri-colored bat.

They all looked alike to me, but Herzog pointed out subtle differences between them: the shapes of their ears, the way some of them gripped the stone with their thumbs as well as their feet. After a few rounds, he asked me to ID a bat on my own. I shrugged a wrong guess.

Herzog forgave me. When he went awhile without looking at bats, he could get confused. "It's a perishable skill, frankly," he said.

We were finding so many bats that we now could see patterns. This mine was full of different set points. Some species clearly preferred to hibernate closer to the mine's opening, where the outside air made it cold and dry. Farther back in the mine, the stagnant air was cool and damp enough to make my glasses fog.

We found a little brown bat that had chosen this set point, its fur turned from brown to silver by the beads of water condensing on it.

When these bats began hibernating a few months before, their torpor freed them from the hard work of staying warm-blooded. Rather than maintain a high body temperature, they could let themselves match the temperature of the air around them. If they were hanging from a tree outside, this strategy would have been suicide. The bitter cold of winter would have frozen them solid, destroying their cells. But inside the cave, insulated by rock and overhanging soil, they enjoyed a chilly but constant temperature. They could, in effect, borrow the mine's own homeostasis.

The bats also had no need for fuel to fly, since they were forsaking hunting till the spring. The females that had mated in the fall were not yet pregnant. They were storing away their sperm till they roused again in the spring, when they would finally fertilize their eggs. Then they could nourish their hungry embryos with fresh food.

Still, the bats hibernating here for the winter were very much alive. They went on inhaling oxygen to burn their ATP molecules. They needed to breathe out carbon dioxide to keep their blood from turning too acidic. With each exhalation, they also let out a tiny bit of water. More water evaporated from their wings. The water they lost in a day wasn't enough to put them in danger. But after two or three weeks the bats felt the homeostatic pinch.

Once they sensed they were running dangerously low on water, the bats took a brief break from hibernation. They warmed back up to their summertime temperature in a matter of minutes. Their reheated bodies allowed them to flit around the hibernaculum and sip water. Now replenished, they could return to a chilly roost to hibernate for another few weeks.

Every time the bats roused, they burned up more of their dwindling supply of fuel. But in the spring, if all went well, they would emerge from hibernation, their homeostatic ledgers still in the black. As I crouched in the mine, though, it was hard to imagine the bats ever coming back to life. They remained weirdly still, hanging alone, in pairs, in a cluster of eleven.

As our tally of bats grew, the mine seemed to me like a crowded zoo. In reality, though, it was a ghost town compared to what I would have seen had I visited in earlier years. When biologists surveyed the mine sixteen years before, in 2004, they counted 1,102 little brown bats. Two years later, things changed. In hibernacula around Albany, biologists found dead bats scattered around the entrances. Some had been scavenged by raccoons. Some had flown into snowbanks. Some had blooms of fungus in their noses. Soon other bat populations around New York crashed, and then other states followed the same steep decline. The dead bats all turned out to be infected by a fungus from Europe called *Pseudogymnoascus destructans.* Its deadly bloom gave its disease the name "white-nose syndrome."

The new disease took over Herzog's life. "It immediately went to the top of the list, and everything else fell far behind," he told me. Herzog watched some species plummet 90 percent in a few years. Others fell 99 percent. Thanks to the decades of records that he and other biologists had made of bats in New York, they had an unrivaled view of the destruction white-nose syndrome caused. "If the disease had arrived somewhere else, we wouldn't have known nearly as much as quickly as we did," Herzog said. He was mourning more than bragging. "I don't know if 'fortuitous' is the right word."

In Europe, the fungus had been harmless to bats. It caused only minor infections that the animals easily kept in check with their immune systems. Somehow the fungus had been transported from Europe to North America—likely to some cave or mine not far from Albany. And somehow it proved deadly to the bats in its new home.

How it killed North American bats was a mystery at first. Pathologists who looked at the dead animals didn't see the kind of overwhelming damage that a lethal fungal infection typically causes. "They were looking at it through clouded lenses," Herzog said.

It gradually became clear that white-nose syndrome is a disease of homeostasis. In the late summer and fall, the bats picked up fungal spores as they visited caves and mines. The cold-loving fungus stayed dormant on their bodies until the bats began to hibernate and their bodies cooled. Once their body temperature fell below about 68 degrees, the spores opened and the fungus pushed threads into their skin and muscle.

Herzog and his fellow scientists discovered that sick bats roused themselves out of hibernation more often than healthy ones. It's possible that they were losing extra water from the sores that formed on their wings. To keep their homeostasis, they had to drink more. It's also possible that the bats were fighting the fungus by warming up their bodies more often, allowing their immune systems to wake up and engage their enemies in brief but intense battles.

Some of the infected bats managed to keep their homeostasis intact until spring, when they could warm up again and fend off the fungus. But others suffered a homeostatic failure, running out of their winter stores. Some became so desperate that they flew out of their hibernacula and into the snowy daytime in a futile search for food. Many were picked off by hawks.

We waded through hip-high water to count more bats, and then we clambered up a loose heap of rocks and grit toward a sliver of light. We crept out into the bright day, the snowstorm having departed off to the east. Ritzko and Herzog compared their numbers, which they would officially log when they got back to their desks in Albany.

The big brown bats turned out to be the most abundant, fifty-four all told. Those numbers hadn't changed in three decades, even after the arrival of white-nose syndrome. For some reason big brown bats were among the few species in New York that didn't seem harmed by *Pseudogymnoascus*, perhaps because they preferred parts of the cave that were too cold for the fungus to grow. The little brown bats, on the other hand, preferred a warmer perch. "They have the worst microclimate," Herzog said. As a result, they were also one of the hardest-hit species. On our 2020 survey, Herzog and Ritzko found only six of them.

It was puzzling that the little brown bats had fallen so far and yet were avoiding outright oblivion. Herzog and his colleagues were exploring the possibility that the few surviving little brown bats carried protective genes. They might have mutations in their DNA that caused them to behave differently in the winter—perhaps preferring a colder perch—in a way that could let them withstand the fungus.

For now, Herzog and Ritzko couldn't do much beyond bear witness. They couldn't try saving the bats by scrubbing a few mines clean of fungus to make them into sanctuaries. The bats themselves would contaminate these refuges by bringing the fungus from the other places they visited. All the scientists could do was observe whether the homeostasis of the bats continued to fail or shifted instead to a new, safe set point.

"We have failed largely to come up with anything we can do," Herzog admitted as we drove out of the woods. "The bats are going to have to figure it out on their own."

COPY/PASTE

One early spring day, I drove to New London, Connecticut, to watch a tree prepare to make more trees. At the north end of the city, I passed through a gate on Williams Street and entered a twenty-acre spread of New England's native trees and shrubs. Its official name was the Connecticut College Arboretum, but everyone there just called it the Arbo. A botanist named Rachel Spicer was waiting for me at the Arbo gate, a borer dangling from her backpack. Whenever she spent time among trees, she brought it along just in case she came across one she wanted to drill. "It's my favorite thing in the world to do," she said.

We wandered down the Laurel Walk, past a tree surgeon who was desperately getting ready for an exam. He stared up at the crown of a Washington hawthorn, checked the species identification app on his phone, and then looked to us, shaking his head in desperation. We wandered farther down the wood-chipped paths, passing American beech and eastern shadbush.

"Sometimes I feel like I was meant to study trees," Spicer said. Her father had taught her how to recognize different species in the forests of Massachusetts where she grew up, and then she went to graduate school for botany, studying the red maple trees of New England and the Douglas firs of Oregon. After she became an assistant professor at Connecticut College in 2010, she set up a lab in which she could study trees up close, growing bits of poplar in petri dishes and inspecting the genes that switched on and off in their cells. The work was fascinating, but Spicer would sometimes get lab fever. When I asked if we could meet, she jumped at the excuse to grab a borer, cross Williams Street, and spend an afternoon with trees in full.

We made our way farther down the sloping gardens to a stretch of low-lying swamp. There we stopped in front of a red maple. It stood like a crooked telephone pole, its canopy high and narrow thanks to decades of competition for light with neighboring trees. A few errant branches shot off the lower parts of the maple's trunk, twisting and turning toward the ground. The branches had been bare for six months, and now it was hard to tell if the tree was even alive. I thought back to the previous summer and tried to picture the maple vibrant with green chlorophyll, capturing sunlight in its leaves to power a molecular machine for making fuel. I flipped forward through the calendar in my mind to autumn, when the chlorophyll in the leaves broke down, the green giving way to red.

"It's not just breaking down because it's old and getting cold," Spicer said. "It's being *deliberately* broken down. Because it is precious."

Spicer explained that every molecule of chlorophyll contains four atoms of nitrogen. If the maple tree in the Arbo simply dropped its leaves in the fall, it would have to make a huge effort in the spring to gather a fresh supply of nitrogen from the soil, which it would then have to pump up from its roots to its branches. Instead, the tree spent the autumn carefully dismantling its chlorophyll into molecular parts, which it moved down little tunnels from the leaves into the branches. There the parts would spend the winter in safekeeping, ready to be quickly moved into new leaves in the spring and reassembled into fresh chlorophyll.

It was a smart strategy but a tricky one. In the summer the thick layer of chlorophyll in the maple's leaves had two jobs: to make food and to serve as a sunscreen. It protected proteins and genes from damage caused by errant high-energy photons. Once the fall arrived and the leaves began breaking apart their chlorophyll, they opened themselves up to attack.

The maple tree defended itself in the most beautiful way possible: it produced a red pigment in its leaves called anthocyanin. The autumn pigment protected the leaves from sun damage for the few weeks they needed to move their chlorophyll into winter storage. Only then did the maple sever its leaves, letting them drop to the ground.

Now in early spring the branches looked dead to me. But the future of the tree was unfolding inside. Spicer grabbed one of the low-reaching boughs and bent it close enough to show it to me at close range.

She pointed to the reddish bulbs that swelled out along its length. After the leaves had dropped in the fall, the branches had produced these bud scales, each coated with a tough outer wall packed with anthocyanins to shield against the winter sun. The bud scales built these defenses to protect delicate new cells inside. These cells were filled with potential: they could become any of the structures that the tree would create in the spring. Spicer used a fingernail to slit open a bud scale. I could see tiny curved streaks inside, some of which might ultimately give this maple tree a chance at eternity.

"There are preformed flowers in there," Spicer said.

Driving home from the Arbo, I looked at the miles of maples flanking the highway. In previous years I didn't pay them much mind in March. But now I was keenly aware of the faint haze of red that floated on their crowns: thousands of bud scales hiding flowers that were coming into being. It looked like a canopy of bloody smoke. Every one of these trees had come into existence out of another red haze on another tree decades ago. They all had ancestors, as do we all, along with every other living thing on Earth.

As a hallmark of life, reproduction is as hard to miss as the screams of childbirth. People make people, maples make maples, dog's vomit makes dog's vomit. For all species, the core of reproduction is the same: the generation of new organisms that carry copies of their forerunner's genes. The details of human reproduction—how cells copy their DNA as they divide, how eggs and sperm end up with only half a set, how they are combined at fertilization, how embryos develop in the womb—is the version we're most familiar with. But it's a mistake to generalize too far from our own species.

What's true for humans is fairly true for another mammal, like the northern long-eared bat. Both species have uteruses and produce live young that suckle milk. But what's true for us is less true for a python, which hatches from an egg, and it's far less true for slime molds like *Physarum polycephalum*.

One way for slime molds to reproduce is by making spores, which can get carried away by wind or water. If the spores of *Physarum* land in a promising spot, they tear open and cells crawl out. Slime mold experts refer to these cells as amoebae. Like our eggs and sperm, each amoeba has only half a set of chromosomes. But despite this shortfall they can live on their own. They crawl across the forest floor, destroying and eating bacteria they encounter. If they happen to bump into another *Physarum* cell, the two can merge together in an underground version of fertilization, to create a slime mold version of an embryo.

Slime mold amoebae are not male and female cells, but they do have their own bizarre version of sex. When two amoebae meet, they inspect the proteins on each other's surfaces. Depending on the versions of these proteins that slime molds inherit, they may belong to one of hundreds of different mating types. As long as two amoebae don't belong to the same mating type, they can merge. Their chromosomes join together into a full set, whereupon the single fused cell becomes a kind of slime mold embryo. It now starts to grow tentacles, making new copies of its chromosomes that populate the single, gigantic cell.

Physarum has yet more strange ways to reproduce. It can, for example, skip the sex and simply dry up into sclerotia. If these fragments blow away in the wind and start growing elsewhere, that single network can turn into many new ones. You can think of the new slime molds as genetically identical offspring. Or you can think of them as just one giant network with some big gaps in it. The slime mold does not care about these word games. It just keeps looking for food.

The alien sex life of slime molds happens largely out of view, decipherable only by the scientists who dedicate their lives to such things. Maple trees, on the other hand, mate across the sky. After my visit with Spicer at the Arbo, I spent the following weeks closely observing the maples in my life. A red maple looms in the far corner of our backyard, and we have a scattering of smaller Norways and silvers. In the salt marshes along the edges of my town, along the verges of the streets, on the flanks of hills, in the empty lots—everywhere, new maples volunteer in endless supply. Through the spring, I watched as one species of maple after another split its buds, putting forward different versions of flowers, some pale green and some crimson. The trees bloomed before leafing, building their flowers solely from the ingredients they had stored away in their twigs the previous autumn.

Like many other plants, maples put forth flowers that botanists call male and female. Those labels stick only lightly to the trees, though, because plant reproduction is so different from ours. A red maple may put out male flowers one year, switch to females the next, and then put out both male and female flowers the year after that. The reason that botanists call a maple tree's flowers male and female is that each kind produces sex cells that follow some of the same rules as eggs and sperm. Just as men produce small sperm, male flowers produce small pollen grains; women make eggs, and female flowers make ovules, which will become seeds after pollination.

Humans come together to have sex, but maple trees need the wind to unite their sex cells. Although they can withstand hurricane winds, even a slight breeze can whisk away their pollen. Most grains land on the ground or the wrong kind of tree. Even if pollen alights on another red maple, it will likely land on the bark or a bough. Only a minuscule fraction of pollen grains have the good fortune to reach a female flower.

The flower snags the pollen on sticky hairs, and a tunnel forms from its surface to its core. The pollen gets pulled through it until it reaches the female flower's ovule. When they fuse, pollen and ovule form a new genome, which gets stored away in a new seed.

I couldn't witness these invisible fertilizations, but I could watch the result: the female maple flowers fell away, leaving behind fleshy red structures that look like pairs of impala horns. These growths, called samaras, held a pair of seeds at their base. The horns grew long and then grew flat. They took on the shape of curved blades, their surfaces resembling stiff paper. When they broke apart and dropped from their stems, they didn't fall so much as fly.

A samara's blade has the same overall geometry as a wing, which it uses for the same purpose: to manipulate the air around it into flight. But while a bat grows wings to catch prey and find a place to hibernate, maple trees grow wings to spread their seeds. The seed at the base of the samara is heavy enough to fall fast, creating a rush of air flowing up along the papery blade. The samara spins like a helicopter, generating lift. The result is a long glide that can carry a maple seed hundreds of feet from its parent tree before it finally reaches the ground.

It took only a few days for each of our maples to drop all its samaras. In a good year a seed rain may bring down nearly a hundred thousand samaras from a single tree. A one-acre stand may shower as many as 8 million. It is a spectacular feat of reproduction. It is also a spectacular waste. As many as half of a maple's samaras are empty, lacking a seed. A sizable fraction of the seeds in the remaining samaras commit suicide. Scientists don't yet understand the evolutionary logic behind these dead and empty vessels. Trees may make hollow samaras as decoys, tricking squirrels and birds into wasting their time and thus giving their seeds a better chance to sprout. Seeds may commit suicide if they happen to carry bad combinations of genes that make it unlikely they'll produce healthy trees.

In the end, only a small fraction of the samaras in a seed rain manage to sprout. Yet, even after this decimation, the trees are still left with an absurd number of progeny—sometimes dozens of viable seeds on every square yard below them. They need little sunlight and not even much soil to sprout roots and send up shoots.

As the spring progressed, samaras carpeted the grass. I live on half an acre of grass-covered pink granite, with plenty of bare patches where the ancient volcanic rock pokes out. Volunteer maples unfurled fingernail-sized leaves. I climbed a ladder to muck out fistfuls of samaras from the gutters. I even found seedlings growing there, as if they could start an aerial forest.

In the summer, my wife and I drove out into the forests that surround our town to hike. One day we traveled through a maple stand where the ground was covered in a great green shallow lake of foot-high saplings. Only a few pole-stage maples rose above them. Fewer mature maples were growing toward the light. And even fewer ancient trees towered over them all, spreading their branches out to form a canopy.

Here we could see the dismal odds in the life of a maple tree laid out before us. As a hallmark of life, reproduction is not as simple as the others. Every living thing metabolizes food, makes adaptive decisions, and keeps itself in homeostasis. The alternative is death. Every living thing is the result of reproduction, but it is not guaranteed to reproduce. If a maple tree lives for its full life span—over a century for some species, three centuries for others—it may rain millions of flying offspring. But only a few will ever manage to rise to meet them crown to crown. The unconscious competition continues on through the generations. A maple tree may succeed in passing down its genes to a few progeny, only to have them all die of root rot.

The maple trees that rain samaras on us today have a deep genealogy. Maples arose more than 60 million years ago, not long after an asteroid crashed into the earth and wiped out the giant dinosaurs. They originated in East Asia, where species like the Nippon maple and lime-leaved maple still grow, and by about 30 million years ago, maples had spread their samaras to North America. They continued to diversify into new forms. The red maple and silver maple that grow side by side in backyards today are distant cousins, having diverged from a common ancestor 10 million years ago. These looming trees are the product of a thin pedigree of success that has sliced through a vast field of reproductive failure.

In fact, it is the intertwined failure and success of maple tree reproduction that has produced all their impressive adaptations—their keen sense of the calendar, their sunscreen, their living helicopters. It is responsible for their diversity, too, having generated 152 species of maple trees all told. Out of life's success and failure at making copies of itself emerges its most impressive hallmark of all: evolution.

DARWIN'S LUNG

The petri dishes were stacked high, like a laboratory pillar. The top dish had a cerulean coat, the color of the sky just after sunset. The dishes below were blue as well, but the farther I scanned down, the paler they became. By the time my eye reached the ground floor of the tower, it had become transparent.

I encountered this plastic monument in Osborn Memorial Laboratories, a castle-like building on the campus of Yale University. It had been stacked by a researcher named Isabel Ott. She had short jet hair; her earrings were coaster-sized disks decorated with all the phases of the moon. Ott had graduated the year before from the University of Georgia, where she had studied all manner of diseases, both human and animal. She had come to New Haven to work for an evolutionary biologist named Paul Turner. For Ott, stacking the petri dishes was not a laboratory version of Jenga. It was the start of a day's work that might eventually save someone's life.

The dishes, Ott explained to me, got their blue cast from the bacteria growing on them. And those bacteria had come from the lungs of desperately ill people who were running out of hope. Some of the volunteers sent Ott notes along with their samples, writing of their plight and begging her for help. "They're my age," Ott said. "And I say, 'Sorry. I am trying everything.'"

The people who had provided Turner and Ott with these bacteria all had a broken gene. Normally, lung cells use the gene to make a protein called CFTR that helps keep the airway clear. But mutations in the CFTR gene disable the protein. The lungs of people who inherit this mutation become obstructed with a thick, sticky layer of mucus.

The disease is known as cystic fibrosis. One of the most dangerous results of this disease is that the lungs become an incubator for certain kinds of bacteria. A species called *Pseudomonas aeruginosa* poses one of the biggest dangers. Normally it lives on the leaves of plants and on clumps of soil. If healthy people happen to breathe in a snort of *Pseudomonas aeruginosa*, their immune systems quickly wipe it out. But the congested airways created by cystic fibrosis give the bacteria refuge, a chance to take hold. Half of people with cystic fibrosis are colonized by an infection of *Pseudomonas aeruginosa* by age three, and 70 percent of adults will develop a chronic infection. Antibiotics sometimes kill off the bacteria but often fail. As the years pass, the bacteria cause inflammation and scarring that make it harder to breathe.

Ott was helping run an experiment that might offer a new way to attack the bacteria. She and her colleagues were testing their idea on volunteers with cystic fibrosis. To see how effective their attacks were, the participants periodically coughed mucus into tubes, which they sent to the scientists. The bacteria in their mucus were now growing in Ott's dishes.

If the idea proved valid, the scientists might be able to transform the bacteria from potential killers into harmless nuisances. To carry out this alchemy, they were taking advantage of life's never-ending power to evolve.

Every living thing on Earth is the product of evolution, a process that has unfolded for some 4 billion years. Bacteria and other microbes were the first lineages to evolve, and about 2 billion years ago they were joined by a new form of life. Amoeba-like single-celled organisms began hunting the microbes. Their cells were far bigger, and they kept their DNA stuffed in a sac called a nucleus. These new forms of life were known as eukaryotes.

Today, slime molds and many other species of eukaryotes do very well as single cells. But some lineages of eukaryotes evolved many-celled bodies. Green algae moved ashore about half a billion years ago, becoming mosses and ferns, with flowering plants arising hundreds of millions of years later. Animals evolved from single-celled eukaryotes in the ocean about 700 million years ago, and some of their descendants later crawled ashore—first millipedes and primitive scorpions and other invertebrates, then four-legged salamander-like creatures. Some of the four-legged animals lost their limbs and became snakes. Some modified their legs and began to fly, becoming birds and bats. One lineage of primates stood upright about 7 million years ago and eventually spread from African savannas across the planet, looked back through time, and recognized for the first time the crude outlines of evolution's deep history.

Life continues to evolve today. It can no more escape evolution than water can escape being wet. When a maple tree showers the earth with samaras, it spreads copies of its genes. But each new sapling is not a perfect replica of its parents. It inherits shuffled samples of their chromosomes. Its genes contain new variations. Charged atoms and high-energy photons crash into genes and alter their sequence. When enzymes create new copies of DNA, they sometimes accidentally put a G where there should be a C. Sometimes they accidentally duplicate thousands of bases in a row.

Cells have special enzymes to proofread these mistakes, but

some get through. The egg and sperm that produced you carried mutations that your parents were not born with. New mutations get passed down through the generations along with old ones, building up genetic diversity over the centuries.

Many mutations have no effect one way or another. Some are devastating, causing lethal disorders and deformities. Others are beneficial, helping organisms survive and reproduce. As mutations get passed down, some become more common, and some less. Chance can steer their fate, but if a mutation has a strong effect on how many offspring an organism has, it can meet that fate much faster. As beneficial mutations accumulate in a lineage, they can give rise to new adaptations.

The basic logic behind evolution is simple enough. It's so simple, in fact, that Charles Darwin worked it out in the mid-1800s, decades before scientists recognized genes, let alone figured out what genes are made of. It was enough for Darwin to observe that animals and plants had variations in each generation and that some of those variations could be inherited. He hypothesized that a process he called natural selection would favor the variations that helped with surviving and having offspring.

Darwin could see the results of evolution in living species. But he believed that life evolved the way mountains grew—over millions of years, on a scale that humans were not equipped to perceive.

"We see nothing of these slow changes in progress, until the hand of time has marked the long lapses of ages, and then so imperfect is our view into the long past geological ages, that we only see that the forms of life are now different from what they were," he declared.

Darwin was wrong, but he can be forgiven. He had no way to appreciate that microbes could display evolution in action over a matter of weeks. They first began to reveal its secrets on the morning of February 15, 1988, in Irvine, California. A microbiologist named Richard Lenski started an experiment with bacteria that would last for decades.

Bacteria can divide in as little as twenty minutes, which means that a single microbe can give rise to a population of billions overnight. Some of those descendants will carry new mutations, which affect how quickly the microbes grow and reproduce. A billion birds require a continent. A billion microbes need only a flask.

Lenski came up with an experiment he hoped would lead to evolutionary change he could measure. He began with a single *Escherichia coli*, a species of gut bacteria that has become the workhorse of microbiology labs. He grew a colony from that founder and divided its descendants among twelve flasks. Each flask contained only enough sugar to last the bacteria a few hours. After their supply ran out, they had to survive till the next morning. Then Lenski and his students drew a little liquid from each flask and squirted it into fresh ones. The bacteria that managed to make the journey could feast on sugar and reproduce again.

To track their history, Lenski created what he liked to call a frozen fossil record. Every five hundred generations his team drew off some of the liquid from each flask and put it in a freezer. Later he could revive them and see how they measured up to their descendants. By the time Lenski moved his flasks and freezers to a new job at Michigan State University in 1991, his twelve colonies had passed through thousands of generations. And they had clearly evolved.

The descendants in all twelve flasks gained mutations that enabled them to thrive in their new environment, taking up food quickly and then surviving their daily fasts. They needed less time to get big enough to split in two. As they gained more beneficial mutations, they kept improving until they were growing 75 percent faster than their ancestors. The microbes swelled along the way, evolving to be twice as big as their ancestors—for reasons that Lenski and his students have yet to figure out. In later years they identified many of the mutations that arose in each of the twelve lines. Different mutations were responsible for some of the changes. But natural selection pushed all twelve lines in the same general direction.

Scores of graduate students passed through Lenski's lab over the years. They tended to his bacteria, became experimental evolutionists in their own rights, and established labs across the United States. Paul Turner was one. Another was named Vaughn Cooper. At a scientific conference in New Hampshire in 2019, I watched Cooper, lanky and eager, give a talk about how high school students could also see evolution in action. He and his colleagues at the University of Pittsburgh had put together a kit that the teenagers could use to run a weeklong experiment. Thousands of students had already gotten results, Cooper announced. Surely, I thought, a writer in his fifties could follow in their wake.

Cooper agreed to send me a kit, and one day a cardboard box showed up outside my front door. I slit open the top to check the contents. I found petri dishes, sealed tubes, bottles of clear liquids, and a bag of black and white beads. The dishes were pimpled with ghostly streaks and swipes. They gave off an unpleasantly sweet odor, like what you might sniff from a jug of cider forgotten for a few days on a picnic table. The dishes were seeded with another species of *Pseudomonas*. *Pseudomonas fluorescens*, which also grows on plants and in the soil, does not attack the lungs of people with cystic fibrosis. It's harmless enough for high school students to safely handle.

I taped the box back shut and put it in my refrigerator, hoping the bacteria's odor wouldn't penetrate the food nearby. I still needed a stand-in for a teacher. I lived near Yale, where I taught a writing class, and Turner and Ott kindly agreed to help me run an experiment.

One day I brought my box to Turner's high-ceilinged lab, where a flock of grad students were centrifuging tubes, spreading microbes on plates, and labeling lids. I set down my box on a space Ott had cleared next to hers. She had me don gray gloves and a white coat. When we took the dishes out of my box, the odor leaped out, too.

"*Pseudomonas*," Ott groaned, like meeting an old nemesis. "If I'm in a bad spot working on it, it can nearly give me a migraine. I have to go and sit on a couch for a few minutes to have a cup of tea so I can get my brain to cooperate."

Before coming to the lab, I had taken an online class on basic lab safety: flushing out eyes, cleaning up spills, and the like. But now, working with Ott, I felt like I was in hygiene boot camp. She instructed me to wipe down my lab bench with alcohol and then light my Bunsen burner. She put her hands around the flame as if she was holding a globe.

"This is your zone of sterility," she announced.

As long as I worked in my zone, I would be safe from all the invisible life that could throw off my experiment. The bacteria and the fungal spores drifting through the air would get torched before they could drop into my tubes and plates and outcompete my *Pseudomonas*.

I labeled a set of plastic test tubes with letters and numbers. Taking up a pipette, I squirted liquid into the tubes laced with nutrients that *Pseudomonas* needs to grow. When I accidentally brushed the tip of my pipette across the counter, Ott had me stop and replace it with a new one. It might have picked up microbes that had rained down on the surface since I had sterilized it, and they might run riot in my tubes.

"There's a spectrum of paranoia," Ott said, "and I'm at one end."

Once I had filled all the tubes, Ott had me douse a pair of forceps in alcohol and then stick them in the Bunsen burner flame. The alcohol flared blue and then died away. I slipped the sterile forceps into the bead pouch, drew out beads one at a time, and dropped them into the tubes.

Now it was time to add bacteria to the tubes. Ott handed me an inoculation loop: a long, stiff wire that ended in a barely visible circle. I scorched it in the flame, lifted the lid of one of the dishes, and scooped up a pinhead bump. The tip of my loop now carried millions of genetically identical members of a strain called *Pseudomonas fluorescens* SBW25, which scientists originally isolated from a sugar beet on an English farm.

I dunked the bacteria into a tube, scorched the loop again, and seeded another tube. When I had filled them all with bacteria, Ott slotted them all in a tray, which she set on a platform in a refrigerator-sized incubator. She turned a switch and the platform began to swivel, sloshing the liquid in the tubes.

The next day the tubes had turned from clear to cloudy, thanks to all the billions of new bacteria that had grown overnight. Even more heartening was the sight of the beads. *Pseudomonas* had covered them in a slimy coat.

To microbiologists, this slime is an architectural marvel. When *Pseudomonas* lands on a surface, proteins on its membrane register the event. They change shape, and that transformation causes them to alter proteins swimming around on the inside of the microbe. A rolling cascade of molecular flips occurs, ultimately causing the microbe to switch on a set of genes. It makes proteins from those genes that it then ferries up to its membrane and squirts out. These proteins weave together into a gooey, sticky matrix. The microbe nuzzles itself into the goo to anchor itself to a surface. It can then feed on the protein fragments passing by. When it grows and divides, its daughter cells release their own goo, spreading their collective slime. Every species of *Pseudomonas* builds biofilms as a way to colonize surfaces—a leaf, a grain of soil, the gut of a grasshopper, the lung of a person with cystic fibrosis.

I moved the slimy beads into new tubes into which I had placed fresh beads. The next day I found that the new beads had become slimy with bacteria, too, and these I moved onward to new tubes. With each transfer of a bead I was playing the part of natural selection.

Every time a *Pseudomonas* divides, there's about a one-in-a-thousand chance it will make a mistake, leaving behind a mutation in a daughter cell. With each cell able to produce a billion descendants in a day, my tubes produced many millions of mutants. And by moving beads from one tube to the next, I was favoring the mutants that were better at forming biofilms on the beads. Any bacteria still left floating in the broth were doomed to destruction when Ott and I sterilized the old tubes.

A week after my first visit, I came into the lab to see if life had evolved under my watch. Ott held up a pair of petri dishes.

"So, basically, here is a normal," she said, moving one dish forward. "And here's your boy," she said, indicating the other. "If you want to glove on, I can take a picture of you and your evolved mutants."

I put on my gloves and held up the dishes with a grin for the iPhone. In one hand I held the normal boy, a petri dish of ordinary *Pseudomonas fluorescens*. It contained bacteria that we had allowed to grow under ordinary conditions for the week. When Ott spread them across a petri dish, they grew into tiny pimple-sized colonies, just as their ancestors had.

In the other hand, I held my boy: my collection of evolved mutants. After I had transferred slimy beads from tube to tube for a week, I put the final tube on a shaker to pull the biofilm off its bead. I spread the bacteria across a fresh petri dish and let it grow into colonies. Ott spied some weird-looking ones, and she scooped one up to seed a fresh dish of its own. There the mutant bacteria grew into dozens of big, fuzzy-edged blobs that looked like ghostly petals on a flower.

Later Ott sent some of my mutants back to Pittsburgh so that Cooper and his colleagues could take a look at them for themselves. They split some of the bacteria open to read their DNA and search for the change that had made them grow strangely.

"It's a new mutant for us," Cooper later told me. The genome of *Pseudomonas fluorescens* contains 6.7 million base pairs. Printed out, it would run about as long as the entire Harry Potter series. In all that DNA, Cooper and his fellow scientists found two genetic typos new to my bacteria. The researchers suspected that one of those mutations—a C flipped to a T—was responsible for the strange flower-shaped colonies that the bacteria formed. The mutation altered a gene that normally helps weave a cotton-candy-like shroud of sugar around each microbe. The mutation in my bacteria likely made that shroud stickier, making them better able to cling to beads and to each other.

Cooper and his graduate students have run high-powered versions of my experiment, moving slimy beads from tube to tube for months. By letting evolution work longer on their microbes, they've created an impressive menagerie of mutants. Some grow into colonies that look like a splash of ink. Others look like slices of kumquats. Some take on the color of tangerines, others of blood.

These colors and shapes are probably just side effects of mutations that make the bacteria grow better in biofilms. The diversity of the colors and shapes Cooper's team has created may reflect the jungle-like complexity of life in a biofilm, a place in which evolution can fill many different niches with many different mutants.

Over the week I had spent in the lab, Ott was conducting an experiment of her own next to me. I could see she was handling a tremendous number of tubes, flasks, and dishes, but I was so preoccupied with grabbing slimy beads with tweezers without shooting them across the lab that I couldn't ask her much about it. Once I had successfully reared a mutant, I asked Ott to tell me more about the cerulean tower.

Her experiment was part of a project to understand how *Pseudomonas aeruginosa* evolves in the lung. A human body is radically different from a leaf or a pond, meaning that when the bacteria first slip into a host, they are poorly suited to their new home. They divide slowly at first. When mutations arise, some of them help the bacteria fare better in the human lung, and they grow faster. Mutation after mutation accumulates. They make biofilms well suited to the airways. If doctors douse them with antibiotics, new mutations can arise that help them shut out the poison. For the bacteria, human lungs are like Lenski's flasks.

Ott was working on a study to find a way to seize control of this evolution. Rather than try to kill the bacteria, the researchers wanted to render them harmless. They would do so by running a version of the experiment I had just carried out. They would alter the bacteria's environment, driving natural selection in a new direction.

The blue color in Ott's dishes came from a pigment produced by the bacteria. Known as pyocyanin, it's the hallmark of

Pseudomonas aeruginosa infections. Indeed, when doctors first isolated the bacteria from sick patients in the late 1800s, they called the microbes "bacteria of blue pus."

Many decades later, scientists began learning what pyocyanin actually does. Among other things, it appears to ward off immune cells that would otherwise attack the bacteria. But it also helps trigger the inflammation that causes so much damage in the lungs of people with cystic fibrosis. If the bacteria simply stopped making pyocyanin, they would become much less of a threat. A researcher in Turner's lab named Benjamin Chan found what might be a tool to push the evolution of *Pseudomonas aeruginosa* in that direction: a virus.

Viruses that infect bacteria are known as bacteriophages, or phages for short. Each strain of phage has molecular hooks that can grab onto a particular kind of protein on the surface of bacteria. Once they latch on, they can invade the microbe and make new phages inside.

Bacteria have evolved a number of defenses against phages, and they can evolve new ones when they come up against a new enemy. One of the simplest ways to protect themselves against a phage is to lose the protein it grabs. If the phage uses a key to unlock its way into bacteria, the bacteria can just get rid of the door. Of course, bacteria put proteins on their surface for a reason. They use some of them to pull in nutrients, others to send signals to their fellow bacteria, and others to act as sensors to tell them about their environment. But the cost of losing one of these proteins may be smaller than the benefit that comes from a defense against phages.

Searching for new species of phages, Chan found dozens that infect *Pseudomonas aeruginosa*. When he and his colleagues unleashed one of the phages on the bacteria, natural selection favored mutants that stopped making the protein the phages used to get in. But the mutation had another effect: it also caused the bacteria to make less pyocyanin. It's possible that the mutation shuts down a genetic switch in the bacteria's DNA that controls both the surface protein and the blue pigment.

Chan and his colleagues wondered if they could use their newfound phage to help people with cystic fibrosis. If they inhaled the phage, it would be unlikely to harm them because phages only infect bacteria, not our own cells. The bacteria might evolve resistance to the phage, but in the process, they would sacrifice their ability to make their dangerous blue pigment.

In a clinical trial, doctors sprayed Chan's phages into the airways of people with cystic fibrosis. As the phages began to attack their *Pseudomonas aeruginosa* colonies, the volunteers coughed up sputum into tubes from time to time. Those tubes made their way to Ott, who isolated the *Pseudomonas aeruginosa* and spread them on petri dishes. After the bacteria grew into a microbial lawn, she stacked them into a tower. The blue dishes on the top came from patients before their phage therapy, and the ones below came from samples taken afterward, week after week. As the blue faded further down the tower, Ott and her colleagues could see that their evolutionary hunch was right. The bacteria were steadily giving up pyocyanin, perhaps becoming safer residents of people's lungs.

"If there's less blue, there's less inflammation," Ott said, "which is a good thing."

Ott and her colleagues would need to take a close look at the bacteria—and run their experiment many more times—before they could determine if the phages were a safe and effective way to tame the microbes. What government regulators might say to the idea of infusing viruses into sick people was anyone's guess.

But sooner or later, with enough persistence, this living medicine would probably work. Evolution made it hard to deny.

Biology expands our vision of life, letting us see beyond our own experiences of being alive, look back over billions of years of living history, and peer down into the microscopic confines of a cell. But every biologist faces a harsh trade-off. No one can know everything about everything. To become an expert on just one kind of life can demand an entire career. Isabel Ott could regale me with tales of disease-causing bacteria if I asked. If I were to quiz her on pythons instead, she would have little to say. I spent hours in conversation with Stephen Secor about pythons as we sampled some of Tuscaloosa's finest microbrews. But I would not go to him to understand the reproductive biology of maple trees.

Yet maple trees and snakes and *Pseudomonas* and slime molds and bats are all joined together by their hallmarks. They all reproduce and evolve; they all make decisions, turn food into energy, and maintain an internal balance. People knew a little about some of these hallmarks before the rise of biology. They knew that trees produced seeds that became more trees. They knew that bats somehow managed to live through both the cold of winter and the heat of summer. Now they know much more about why such things are so. And they find that what is true for one species is, in a fundamental sense, true for all. From time to time, researchers have wondered what these different strands of unity together create. If all living things share certain hallmarks, can they tell us what life is? That question—What is life?—may seem like it's the first and foremost question biologists should answer. And yet it remains unanswered and, perhaps, ultimately unanswerable.

PART THREE A SERIES OF DARK QUESTIONS

THIS ASTONISHING MULTIPLICATION

Waves crashed onto the beach, delivering fresh sand with each break. The beach rose into lines of dunes, like waves of land. On their lee side they surrendered to an orderly landscape of living things: topiaries, parterres, and orangeries. The sprawling estate, known as Sorghvliet, was the summer residence of Count William Bentinck of the Netherlands. In the eighteenth century, it was considered one of the most delightful gardens in Europe. Today little of Sorghvliet's former glory survives: gone is the giant artificial hill enclosed by a maze; gone are the trees with doorways and windows cut out of their branches to make them look like houses of leaves.

To biologists, however, Sorghvliet remains a sacred place. For them its glory does not lie in its once-magnificent grounds but in a tiny animal that lurks in its fishponds and canals. It came to light in the summer of 1740, when scholars across Europe were confidently making declarations about what it means to be alive. And this mystifying creature revealed how deep their ignorance of life truly was.

That animal was discovered by a rootless young man named Abraham Trembley. Trembley had officially come to Sorghvliet to serve as a tutor to the count's two young sons, but soon he was acting as a stand-in parent to the boys. Their mother had left for Germany to live with her lover, and their father spent most of his time at the Hague, either on matters of state or his impending divorce. As isolated as the boys were at Sorghvliet, Trembley was even more so. He had been born in Switzerland, training in mathematics and theology as he prepared to join the church. But political strife drove him to Holland; he scraped by on private lessons for a few years before the count gave him a stable position.

A teacher both pious and curious, Trembley decided he would make it his mission at Sorghvliet to teach the boys to see God's omnipotence in nature's works. Trembley didn't spend much time on formal lessons or regurgitating Aristotle's writings. He was a child of the Scientific Revolution, which had ushered in new theories about life, and he wanted to see how well they explained the natural world with his own eyes.

"Nature must be explained by Nature," Trembley later said, "and not by our own views."

At Sorghvliet, nature was happy to help. Trembley and the boys wandered the grounds to observe animals and plants. They fished out duckweed floating in the ponds. They scooped up insects in the ditches. They brought their hauls back to Trembley's study, where they examined the fine anatomical details of their specimens. Sometimes they used magnifying glasses. Other times they used a custom microscope provided to them by the count, its lens attached to the end of a jointed arm.

Trembley carefully sketched what he and the boys saw in this miniature world. Very often they were the first people ever to see it. Trembley wrote to other scholars across Europe about what they were observing—the strange complexities of caterpillars, bees, and aphids—and his correspondents quickly recognized this teacher, alone on the Dutch coast, as one of their own.

The animals of Sorghvliet soon drew Trembley into a debate about the nature of life itself that had been roiling Europe for almost a century. On one side of the debate were the followers of the seventeenth-century philosopher René Descartes. Descartes attacked traditional notions of nature having goals, such as gravity carrying objects to the middle of the earth as if it knew where that was. Instead, Descartes offered a vision of matter in motion. At first he filled that vision only with inanimate objects like pendulums and planets. But eventually Descartes came to see living things as matter in motion, too. They were made up of parts that worked together, much like a clock. The parts of a clock were set in motion by springs and weights. The parts of an animal's body were likewise set in motion—Descartes believed by tiny explosions inside their nerves. He expected that someday life would be as well described by physicists as a rock dropped to the ground or the moon orbiting the earth.

Descartes inspired generations of followers, and they extended his machine-centered view of life from animals to humans. Aside from our rational souls, they argued, our bodies were much like machines. Cartesian doctors saw themselves in league with clock repairmen. "Like all of nature, medicine must be mechanical," the German physician Friedrich Hoffmann declared in 1695.

But Descartes also inspired generations of opponents. Some were simply appalled that he seemed to have no need of God to explain the world. Others could not reconcile Descartes's vision with their own understanding of nature. The closer these anti-Cartesians looked at life, the more complex it proved to be, both in anatomy and in behavior. And this complexity served a greater purpose: it allowed living things to survive and reproduce. No mechanical philosophy could encompass that complexity or explain its purpose. That purpose, the anti-Cartesians firmly believed, created a decisive difference between inorganic matter and living things.

The physician Georg Ernst Stahl declared that the mission of science was to understand the difference—to get to the bottom of what exactly set life apart. "Above all else, consequently, it comes down to this: to know, what is life?" Stahl said in 1708.

Stahl offered an answer to his own question: a definition that would be followed by many others in the centuries that followed. Life was "not the matter of the body—anatomy, chemistry, the 'mix' of fluids—but rather their interdependence." Stahl believed life's interdependence served the purpose of allowing living things to endure the assaults of a hostile world and resist the forces of decay. There had to be an inner force—what Stahl referred to as a soul—to sustain the interdependence of life.

One of life's most obvious hallmarks was reproduction, but naturalists were fiercely divided over explaining how life managed this feat. Some scholars argued that the parts of a living thing already existed in an egg or sperm. A seed that could sprout into a tree also contained the future seeds of that future tree, they believed, which contained their own seeds in turn. Other scholars found it absurd to imagine that life could exist in such an infinity of boxes within boxes. They argued that the parts of living things did not exist before the things themselves. The complex anatomies of animals and plants must unfold gradually, in a mysterious process of development.

When Trembley began trading letters with other naturalists in 1740, they shared with him a startling discovery about how life can reproduce. From Geneva, Charles Bonnet wrote that he had observed female aphids bearing offspring without mating first. Both Bonnet and his mentor, the French naturalist René Antoine Ferchault de Réaumur in Paris, relayed their astonishment to

Trembley, who decided to investigate the paradox for himself. He and the boys reared some female aphids at Sorghvliet. Just as Réaumur and Bonnet had promised, they eventually began laying eggs.

If female aphids could produce ordinary offspring without sex, their eggs must contain preformed aphids. That possibility hinted that animals could have more autonomy than one would expect from divinely created machines. Trembley wondered if the laws of nature that scholars were claiming might actually be presumptions. His observations made him even more humble in his work. He was content to observe patterns rather than claim to have uncovered God's laws.

That humility allowed Trembley to notice something else that had gone overlooked. One day in June 1740 he was inspecting a duckweed plant that he had collected from a ditch. He noticed a tiny green trunk stuck to its side. Atop the trunk was a strange crown that looked like strands of silk. When he looked at more duckweed, he found more trunks.

Trembley didn't know it, but other naturalists had observed this strange form of life forty years before. They had classified it as a plant. Trembley assumed it was a plant, too, as did visitors to whom he showed it. Some thought he was looking at bits of grass or the tufts of dandelion seeds.

But Trembley noticed something bizarre: their crowns moved. They didn't simply sway in the current of water in his jars. The threads of the crown seemed to move with intention.

"The more I followed the movements of these arms, however, the more it seemed that it had to come from an internal cause," Trembley later recalled.

He grabbed one of the jars containing these strange things and jogged it a little. To his surprise, the threads suddenly pulled back inside the little green trunks. Once he let the jar settle, the threads snaked their way back out again. Their behavior, Trembley said, "roused sharply in my mind the image of an animal."

One day Trembley found that the crowned animals were now stuck to a side of a jar where they had not been before. They were moving on their own, he realized, uprooting themselves from the duckweed and crawling like underwater inchworms. These animals had a goal: over time, they made their way to the light. If Trembley swiveled a jar to put them in the dark, they crept back to the sun. The creatures were also eating. Trembley watched them grab worms in their arms, pulling in their prey to stuff into a mouth nestled at the center of their crown. He observed them eating water fleas and even little fish.

These animals were stranger than anything Trembley had ever seen or read about. He thought up experiment after experiment to make sense of them. He sliced one of the creatures in two, but instead of dying, the pieces regenerated into two full-blown individuals, complete with trunk, head, and tentacles. They even began walking again. "I did not know what to think," Trembley confessed to Réaumur.

Neither did Réaumur. Trembley's letters about the creatures were getting more fantastical. When he told others what Trembley was seeing, they simply refused to accept such a thing could exist. Réaumur asked for some of the creatures to look at for himself.

Trembley packed fifty of his creatures in a glass tube that he stopped with Spanish wax. When Réaumur received the tube in France, the animals were all dead, the wax having suffocated them. Trembley tried again with a corked tube, and one day in March 1741, Réaumur received a live batch of the animals. When he sliced them into pieces, the creatures regenerated just as Trembley had promised. "This is a fact I cannot accustom myself to seeing, after having seen and re-seen it hundreds of times," Réaumur confessed.

If Trembley's animal was an exquisite machine, snapping it in two should have stopped its parts from working. But if animals developed from some preformed seed, regenerating entire new creatures should have been impossible. If every animal was endowed with an indivisible animal soul, did cutting a single individual into pieces lead it to make new souls, without God's foresight and planning?

"Is the soul divisible?" Réaumur wondered.

Réaumur suggested to Trembley that the animals needed a name. He proposed "polyps," adapting the Latin word for octopus. (Today they're known as *Hydra*, and their genes reveal a kinship to jellyfish and corals.) When Réaumur showed off Trembley's polyps to the Academy of Sciences, he inspired in his colleagues the same awe he had originally felt. The official report on his demonstration sounded more like a circus barker's spiel than a scientific paper: "The story of the Phoenix who is reborn from its ashes, fabulous as it is, offers nothing more marvelous than the discovery of which we are going to speak."

With Réaumur's endorsement, the polyp became famous across Europe. Naturalists begged Trembley for animals of their own to look at. "I am entirely taken up with dispatching polyps to one place or another," he groused. When he sent his first batch of polyps to London, two hundred people gathered at the Royal Society to look at them through microscopes. Henry Baker got hold of the Royal Society's polyps, observed them, and drew the animals in their acrobatics. He whipped off a book, *An Attempt Towards a Natural History of the Polype.* While Trembley quietly continued to do experiments at Sorghvliet, Baker satisfied the public's curiosity.

The secondhand stories of the polyps, Baker explained, "have appeared so extraordinary, so contrary to the common Course of Nature and our received Opinions of *Animal-Life*; that many People have look'd upon them as ridiculous Whims and absurd Impossibilities." Baker provided a lyrical firsthand account of how the polyp moved, caught prey, and devoured it. Still, he knew that some skeptics would scoff at the fantastical idea that the polyp was an animal because it happened to be "unsuitable to their Hypothesis of Life in general."

Their regeneration was most unsuitable of all. "If the Animal Soul or Life, say they, be one indivisible Essence, all in all, and all in every Part, how comes it, in this Creature, to endure being divided forty or fifty Times, and still continue to exist and flourish?" Baker asked.

While Baker sang the praises of polyps, Trembley discovered even more extraordinary things about them. Their bodies were held together by a gooey substance that reminded him of an egg white, tenaciously resisting his efforts to pull it apart. Trembley wondered if he was handling the stuff of life, a glue that not only held his animals together but also gave them the power to move. It was the first time anyone had conceived of such a life-giving substance—what scientists a century later would refer to as protoplasm.

In other experiments, Trembley turned single polyps into dozens. He lopped off bits of animals so that they grew back as deformed monsters. He fused two polyps together and found they could live comfortably as a single animal. One day he held a polyp in a drop of water cupped in the palm of his hand. With his other hand, he snaked a boar bristle into the trunk of the animal and drew it out. The animal's body was reversed, like a glove quickly shucked off at the end of a winter day. Its interior now its exterior, the polyp lived on. When Trembley reported this feat, many naturalists considered it impossible. He had to round up a group of prominent experts to gather at Sorghvliet to watch him turn polyps inside out and serve as his witnesses.

In 1744, Trembley finally published a two-volume monograph about all this work. But far from launching his new career as a zoologist, it marked the end of his scientific research. The Bentinck boys were growing up and no longer needed his services. During Trembley's time at Sorghvliet, the count had introduced him to a powerful network of acquaintances who recognized his sharp intellect. He joined a secret diplomatic mission to France to settle the War of the Austrian Succession. When that task was finished, Trembley was hired to oversee the education of a young English duke, traveling the continent with his student for years. Those two posts earned Trembley a pair of lavish pensions, which he used to return to Geneva, buy a mansion, raise five children of his own, and write a series of books on teaching.

Trembley's most important student had been himself. Over the course of only four years, he had taught himself how to run rigorous experiments on animals. As he learned how to extract knowledge from his polyps, he invented the science of experimental zoology. Long after he had finished his work and left Sorghvliet, the discoveries he had made in his polyps haunted the minds of naturalists and philosophers. The creatures—which some incorrectly called insects—demonstrated that life was different from what anyone had previously thought.

"A miserable insect has just shown itself to the world and has changed what up to now we have believed to be the immutable order of nature," the naturalist Gilles Bazin declared. "The philosophers have been frightened, a poet told us that death itself has grown pale."

IRRITATIONS

As Trembley and the Bentinck boys splashed in ditches, a young doctor was building a more conventional sort of fame in the German town of Göttingen. Albrecht von Haller had moved there in 1736, lured by a local baron who was building a new university and needed the best anatomist in Europe to staff it. At twenty-eight, Haller was already the obvious choice. The baron wanted him so badly that he built Haller a mansion, a botanical garden, and a Calvinist church for him to worship in. But that was not all the baron would give Haller. "Called to Göttingen, there was nothing more important for me to do than to build an anatomical theater," Haller later wrote, "and to supply it with corpses."

Like Trembley, Haller was a Swiss in exile. He was born near Bern into a family with a reputation as nervous, secretive, and eccentric. At age five, young Albrecht would sit on the kitchen stove and preach to the family's servants from the Bible. By the time he was nine, he read Greek fluently and had written biographies of more than a thousand famous people. He developed a curiosity about the insides of bodies, which he satisfied by cutting open animals. When he left Switzerland for medical school—first in Germany, then in Holland—he began to cut open humans.

Haller's fellow medical students found him annoying. He took opposing opinions as personal attacks. "He was unable to bear another's error in silence," one biographer wrote. When a famous professor announced he had found a new salivary duct, the teenage Haller carried out an experiment to see it for himself. Haller humiliated the professor by proving he had been fooled by an ordinary blood vessel.

After finishing medical school, Haller traveled to London and Paris to continue his studies. Watching a horrific surgery to fix a bladder, he decided he would not operate on living people. Now he spent even more time with cadavers, and the closer he looked in them, the more things he found. "To go through everything, and to have fully seen all the regions of the human body, is as difficult and as rare as a full account of all the immense districts, rivers, valleys, and hills," Haller later said.

His studies finished, Haller returned to Bern to work as a family doctor, which mostly meant bleeding mothers and children. His meager practice left him with a lot of free time, much of which he spent wandering in the Alps. Still in his early twenties, Haller became famous for his alpine botany. In London he had developed a taste for English poetry, and now he wrote a long, romantic poem, "Die Alpen," in tribute to the mountains. It made Haller the most-read German poet of his day, and it turned the Alps into an eighteenth-century tourist magnet.

When he wasn't bleeding, writing, or hiking, Haller was dissecting—mostly the cadavers of criminals and the poor. He found new muscles, junctures, and vessels. In 1735, Haller carried out the first careful dissection of conjoined twin babies. The babies, who had died shortly after birth, had separate brains but a shared heart. Haller concluded that the soul could not travel in the blood, because that would mean that the two distinct souls of the babies would have to mix together. Far from a deformity,

Haller saw the exquisitely merged anatomy of the twins as further evidence of God's design and omnipotence.

The reputation Haller built in Bern brought him an invitation to the new university in Göttingen, and there he immediately began working even more furiously, publishing volumes on botany and anatomy despite the deaths of two wives and two children over just a few years. In his theater Haller oversaw an army of anatomists as well as artists who sketched everything they revealed in the cadavers.

Dead bodies helped Haller understand living ones. He liked to call the science he was developing "anatomy in action." On the first floor of the theater, Haller ran experiments on human cadavers, and on the second he carried out an even grislier sort of research. There he and his students worked on live dogs, rabbits, and other animals. It was not enough to observe how the domed sheet of the diaphragm was attached to the ribs in a dead man's chest. Haller needed to see living diaphragms in motion.

Trembley carried out gruesome experiments of his own on polyps, but no one was much concerned with the plight of the tiny creatures he cut in half. Haller, by contrast, became notorious around Göttingen for the way he made his animals suffer. "One needs a stock of dogs and rabbits, which have their difficulty in a small town where everything astonishes and attracts gapers," Haller complained.

The pain he caused his animals took a toll on Haller as well. He once described his research as "a species of cruelty for which I felt such a reluctance, as could only be overcome by the desire of contributing to the benefit of mankind, and excused by that motive which induces persons of the most humane temper, to eat every day the flesh of harmless animals without any scruple."

At first Haller designed experiments to understand one organ at a time. Gradually, though, he came to see the diaphragm, the heart, and all the other parts of the body as part of one great system. Haller's mind moved to more fundamental questions about life. For Haller, no question was more important than how living things moved. We can see some of life's motions when we take a walk or blink our eyes. But inside, Haller knew, our bodies are also in constant hidden motion. Our hearts thump. Our gallbladders squeeze out bile. Our intestines ripple.

Haller believed that movements came in just a few forms. Some arise from our will. In other cases we respond automatically to sensations. Haller reasoned that nerves must somehow bring about movements like these. Based on what scholars knew of nerves at the time, Haller believed that they must also sense what happened in the parts of the body they moved.

To see if this was true, Haller and his students probed the interiors of hundreds of living animals with knives, heat, and scalding chemicals. Screams and struggles revealed to them which parts of the body were sensible. The skin, not surprisingly, was exquisitely sensible. But the lungs, hearts, and tendons were not. Probe them as much he might, Haller found no response.

Haller also recognized that nerves were not always required for the body to move. After he removed hearts from animals, the organs sometimes continued to beat long after being severed from the nervous system. After the hearts grew still, Haller could sometimes reanimate them for a spell by touching them with a knife or exposing them to a chemical.

This second kind of movement—known in the eighteenth century as irritability—intrigued Haller even more. He set out on another set of experiments to map it across the body. He and his students probed organs and tissues to see if they contracted in response to a stimulus. Some didn't respond, while others did so weakly. But every muscle proved strongly irritable, and the heart, Haller concluded, was "the most irritable organ of all."

Haller wondered what caused sensibility and irritability. In the 1700s, physicians generally believed that nerves contained a mysterious substance called animal spirits. By some accounts these spirits created chemical explosions to move muscles. But irritability did not rely on the nerves, and so the force driving it must come from elsewhere. That place, Haller decided, was within the muscle fibers themselves, where it was generated independent of the soul.

The more Haller contemplated irritability, the more profound it became. He decided that it was the hallmark of life, providing a clear-cut definition for death: the moment at which the heart lost its irritability. As a force, irritability seemed to Haller as profound as gravity—and also as mysterious. Even a gentle poke to a muscle could trigger an outsized response, seeming to defy standard physics.

In 1752, Haller delivered a series of lectures about his experiments, which he published as a book the following year. Coming out soon after Trembley's work on polyps, Haller's research proved equally provocative. People had wanted to watch polyps regenerate with their own eyes. Now, all over Europe, anatomists wanted to run Haller's experiments for themselves. One visitor to Florence in 1755 wrote, "I saw in all corners limping dogs, on which experiments on the insensibility of the tendons had been made."

Some of the experiments confirmed Haller's, but others failed. Critics attacked him for claiming that so much of the body created its own force, independent of a soul. "The enemies of Mr. Haller have everywhere been of a very great number," one of his students observed. But none of his critics could match Haller's scientific output. The sheer volume crushed his opposition. One French physician simply shrugged in surrender, asking, "What to answer to 1,200 experiments?"

Not long after Haller published his findings, he left Göttingen. He abandoned his mansion, his church, his garden, his theater. Now forty-five, Haller returned to Switzerland hoping to gain political power, but he misjudged his chances, merely landing a job running a saltworks. This left him with plenty of time to write about medicine and botany and to publish 9,000 book reviews. Haller would never dismantle another cadaver. He did not flay another rabbit. The closest Haller came to that sort of research was experimenting on himself.

By the time Haller returned to Switzerland, he had lost the energy that had propelled him over mountains as a young man. He was now a victim of fevers, indigestion, insomnia, and gout. Sensibility took its vengeance on him. Haller began to observe it from within, with intense curiosity. When his gout flared, he would flex the tendon in his big toe and record his sensations. Not once did he feel discomfort—or at least not until he bent his toe so far that his skin began to stretch, "at which point," he later wrote, "the pain became unbearable." To Haller, his unbearable pain was a personal proof that skin was sensible but joints were not—and thus must not contain nerves.

When Haller reached his sixties, he began suffering chronic infections in his bladder that forced him to use opium. He was intimately familiar with the drug. In Göttingen he had grown poppies in his botanic garden, extracted opium from them, and given it to animals to observe its effects. Haller noted that opium made the animals less sensitive. A heavily drugged dog would show no response in its pupils when he put a candle close to its eyes. But when Haller checked the animals for irritability, he observed that the opium had a much weaker effect. It only made the intestines somewhat less irritable, while the heart kept on beating normally. Haller saw these results as more evidence that sensibility and irritability were two fundamentally different things.

After Haller published his findings, a Scottish physician named Robert Whytt declared he was wrong. Whytt had run experiments of his own and found that opium slowed the pulses of his animals. Haller's "candor and love of truth," Whytt said, should make him "readily acknowledge his mistake, as soon as he shall discover it." That was not Haller's way, though. He waved off Whytt's work as inferior science.

Haller's pain grew worse. He slept even less at night, and his joints began to ache with arthritis. Despite his familiarity with opium, he was reluctant to take it himself. He had heard rumors of eastern kingdoms where rampant use of the drug caused "terrible weakness of mind," he said. For a leading figure of the Age of Reason, nothing could be more frightening than unreason.

Haller shared his anxieties in letters with an old friend, the British physician John Pringle. As one of Britain's leading doctors—he would later become personal physician to King George III—Pringle soothed Haller's concerns with medical authority. "The dose is not to be measured by drops or grains, but by what can procure you nights free from pain and such frequent irritations to make water," he assured Haller in 1773.

The opium gave Haller immediate relief, he reported back to Pringle, "with the hush of the winds that soothe the raging sea." Along with poetry, Haller also recorded the experience as a scientist, tracking his pulse, noting his sweats and the quality of his sleep. He checked his pulse before and after each dose. He recorded every urination. He noted his farts. As the weeks passed, he grew addicted and the opium grew less effective. Haller increased his dose to 50 drops, then 60, 70, and eventually 130. The opium now launched Haller into blissful hours that were "joyous and of the highest zeal for activity," he wrote. And they were always followed by a crash.

"The already generally weak physical strength is exhausted even more when the effect of opium subsides," he wrote. "I noticed the very repulsive odor of opium exhalation through the skin; in this odor lay something burnt of unpleasant sensation for the nose."

By 1777, Haller was housebound, obese, and partly blind. But he still greeted a flow of visitors to his home, including Emperor Joseph II, who asked if Haller was still writing poetry. "Indeed no," he reportedly said. "That was the sin of my youth."

Yet Haller still wrote, thanks to his steady supply of opium, including a report on his experiences with the drug. To the end, he looked for evidence that he was right and that Whytt was wrong. Haller found that his pulse rose as the opium eased his pain, and it dropped when the drug wore off. Haller used his addiction to tease apart the nature of irritability and sensibility.

Soon after Haller's report on opium was delivered in a public lecture, he died. He had many biographers, and they all loved to tell the story of his last moment of life. This one, from a 1915 biography, is obviously fake but wonderfully apt:

The fingers of one hand rested upon the diminishing pulse in the other. At length he said, calmly: "It no longer beats—I die."

THE SECT

What Haller and Trembley alike cared about most was observing life. They had little desire to forge sweeping explanations of all the things they observed. Haller believed he would never truly understand irritability, because its true nature was, he said, "concealed beyond the research of the knife and microscope." Beyond that boundary, Haller would not venture. "The vanity of attempting to guide others in paths where we find ourselves in the dark, shows, in my humble opinion, the last degree of arrogance and ignorance," he wrote. God had mysteriously invested muscles with irritability, just as He had lodged gravity inside the earth and moon.

But other naturalists dared to explain life for themselves. The leading naturalist of the day, Georges-Louis Leclerc, Comte de Buffon, argued that life was chemically different from lifeless matter because it was made up of things he called "organic molecules." Buffon had no idea of what any molecule was made of, let alone what distinguished one organic molecule from another. But he was convinced that all living things—be they polyps or people—reproduced in the same way: they assembled organic molecules into copies of themselves.

A polyp and a person were both alive because they were made of these organic molecules, and could faithfully reproduce themselves in new combinations of them. The reason that a polyp and a person were different was that every living species had a unique "internal mold," as Buffon liked to call it. A mold drew in some kinds of organic molecules but not others, creating a distinctive body.

Haller and Trembley didn't enjoy watching others use their work to fertilize their own theories. When Trembley read Buffon's claims, he was aghast. "I confess that I can only consider his system as a hazardous hypothesis," he wrote to Count Bentinck. "He makes the facts on which he builds it, prove too much."

Haller was likewise appalled at how theorizers made sweeping claims about his work on irritability. "Irritability is becoming a sect," he groused. "That is not my fault."

The sect was made up of philosophers, naturalists, and physicians who believed that life contains some sort of vital force. These so-called vitalists carried on the fight against Descartes, despite the many victories that his mechanical vision scored in the eighteenth century. Inventors built steamboats, air compressors, power looms, and other devices that would make the industrial revolution possible. Astronomers who treated nature as matter in motion made new discoveries of their own, such as the planet Uranus. But the vitalists pushed back, arguing that life was fundamentally different from a planet or a steamboat. The vital force endowed matter with self-directed motion and the power to generate new complex bodies. The vitalists saw life infused with purpose: eyes were made for seeing, wings for flying, bodies for reproducing. To them, Haller's irritability and Trembley's regeneration were potent examples of what the vital force could do—and what a mechanical view of nature could never explain.

After Haller's death, the vitalists rose to even greater influence. In 1781 the German naturalist Johann Friedrich Blumenbach declared that within all living things "lies a special, innate effective drive, active lifelong, initially to infer their definite form, then to preserve it, and, if it is injured, where possible to reproduce it." Some envisioned the force being passed down from one generation to the next, changing over time to produce different forms.

A British doctor named Erasmus Darwin was the first to share this private notion—what later came to be known as evolution—with the public at large. Today Erasmus is best known as Charles's grandfather, but in the late eighteenth century he was a towering figure in his own right. He wrote a two-volume book that classified every disease known at the time. He made major advances in science as a hobby, developing the first good ideas of how plants use sunlight and air to grow.

Erasmus Darwin believed all his ideas held together in a unified vision of life. He wanted the world to see it, but he knew that most people would not read a dense monograph. And so he created a genre all his own: scientific poetry. Darwin turned the finer points of botany into hugely popular verse. In the age of Wordsworth, Byron, and Shelley, it was Darwin who in the 1790s was the most famous poet in Britain. Samuel Taylor Coleridge called him "the most original-minded man."

Not long before his death in 1802, Erasmus Darwin wrote a poem called "The Temple of Nature." He followed life from its beginnings to the present day.

Organic life beneath the shoreless waves
Was born and nurs'd in ocean's pearly caves;
First forms minute, unseen by spheric glass,
Move on the mud, or pierce the watery mass;
These, as successive generations bloom,
New powers acquire and larger limbs assume;

Whence countless groups of vegetation spring,
And breathing realms of fin and feet and wing.

When the poem appeared a year after Erasmus Darwin's death, it shocked pious readers. Darwin was rejecting the belief that God breathed species into existence in their current form. The reviews of "The Temple of Nature" were savage. One nameless critic sneered at Darwin's "unreal and unintelligible philosophy." It was so appalling that he practically threw his quill away: "We are full of horror, and will write no more."

For Romantic writers like Percy Shelley, however, Darwin's poetry was kindling for literary fires. In the summer of 1816, Shelley and his eighteen-year-old lover Mary Wollstonecraft Godwin—soon to become his wife, Mary Shelley—paid a visit to Switzerland. They stayed with Lord Byron for part of the summer, which was so cold and rainy that they remained cooped up indoors for days at a stretch. To pass the time, they wrote ghost stories for each other.

"*Have you thought of a story?* I was asked each morning, and each morning I was forced to reply with a mortifying negative," Mary Shelley later wrote.

One night the conversation wound its way into "the nature of the principle of life," she later recalled. She listened to her fiancé and Byron discuss Erasmus Darwin's claim that simple life-forms arose from organic matter. They wondered if that meant a corpse could be reanimated. "Perhaps the component parts of a creature might be manufactured, brought together, and endued with vital warmth," Shelley wrote.

Late at night the party ended. When Shelley fell asleep, her mind was flooded with images. She saw a man kneeling beside a stitched-together corpse. He used "some powerful engine," as she described it, to bring the corpse to life, making it stir with "an uneasy, half vital motion." The man then went to bed, hoping the "slight spark of life" in the corpse would go out. But he was then awakened. "Behold the horrid thing stands at his bedside, opening his curtains, and looking on him with yellow, watery, but speculative eyes," Shelley wrote.

She woke up herself. "I have found it! What terrified me will terrify others," she wrote. Shelley eventually turned her ghost story into a full-blown novel, which she published anonymously in 1818. She called it *Frankenstein*.

Her hero, the young scientist Victor Frankenstein, grows obsessed with a question: "Whence, I often asked myself, did the principle of life proceed?" He echoes the language of vitalists, and he follows the example of Xavier Bichat, studying death to understand life. "The dissecting room and the slaughter-house furnished many of my materials," he says.

Before long, Frankenstein has solved the mystery. "After days and nights of incredible labour and fatigue, I succeeded in discovering the cause of generation and life; nay, more, I became myself capable of bestowing animation upon lifeless matter," he declares. Shelley is delightfully cryptic about his success, but she hints that it somehow involves electricity. By the early 1800s it was clear that electricity had something to do with life—a shock could make the legs of dead frogs twitch. But it was still mysterious enough that it could stand in for the vital force of life.

Erasmus Darwin wrote of the vital force with lyrical poetry, describing it like the blooming of a cosmic flower. But Shelley saw something grotesque in science's obsession with life, something that seemed more like an urge to control and exploit. "When I found so astonishing a power placed within my hands, I hesitated a long time concerning the manner in which I should employ it,"

Frankenstein says. He decides to create a living thing by assembling parts from human cadavers. "I collected the instruments of life around me, that I might infuse a spark of being into the lifeless thing that lay at my feet." What he creates might be called life, but only a monstrous kind.

Along with his experiments with electricity, Frankenstein also uses "chemical instruments." Shelley never described exactly what sort of chemistry he carried out, but the very mention of that science would have given a bracingly modern feel to the book. At the dawn of the nineteenth century, chemists were sweeping away the occult mysteries of alchemy, replacing them with elements and atoms.

To appreciate just how revolutionary this change was, consider water. In the 1500s, alchemists tried to define water by its qualities—its transparency, its ability to dissolve substances, and so on—and they ended up in a muddle. Their research revealed to them different kinds of water that shared some qualities in common, but not others. Unlike ordinary water, strong water (*aqua fortis*) dissolved most metals. But only noble water (*aqua regia*) could dissolve the noble metals of gold and platinum.

In the late eighteenth century, the French chemist Antoine Lavoisier demonstrated that water contains molecules made of two atoms of hydrogen and one oxygen. Strong water proved to be no water at all, but instead a combination of nitrogen, hydrogen, and oxygen. Today it's known as nitric acid. And noble water was something else entirely: a mix of nitric acid and hydrochloric acid.

Living things could be broken down into elements as well. But the molecules into which these elements combined in life were hard to find in lifeless matter. Many chemists came to see a vitalist gulf between the organic and the inorganic. "In living Nature the elements seem to obey entirely different laws than they do in the dead," an 1827 chemistry textbook declared.

A chemist named Friedrich Wöhler soon showed that the textbook was wrong, and he used his own urine to prove it. Wöhler experimented with a poisonous acid called cyanogen, mixing it with ammonia. He ended up with peculiar white crystals made from carbon, nitrogen, hydrogen, and oxygen. The proportion of the elements in Wöhler's crystals was identical to those in a molecule called urea, which had only been found up till then in urine.

Our kidneys make urea as a way to pull extra nitrogen out of our blood and flush it out of the body. Chemists in the 1700s first discovered the compound when they let urine evaporate and form crystals. To make sense of his artificial crystals, Wöhler collected his own urine and isolated urea from it. He compared his natural urea crystals to the artificial ones he had concocted from ammonia and cyanogen. Chemically they behaved the same way.

"I can no longer, as it were, hold back my chemical urine," he declared, "and I have to let out that I can make urea without needing a kidney, whether of man or dog."

Wöhler hadn't created a Frankenstein's monster, but he had managed to create an organic molecule without relying on life's vital force. When he published his experiment in 1828, many chemists refused to acknowledge what Wöhler had accomplished. Creating urea from scratch was not all that important, they argued, because it was just one of life's waste products. They continued to argue that only a vital force could create life's organic molecules.

But some researchers followed up on Wöhler's experiment with ones of their own. The German chemist Hermann Kolbe studied acetic acid, which could only be found at the time in vinegar from fermenting fruit. Kolbe discovered how to make acetic acid in his lab out of carbon disulfide, an inorganic molecule produced from coal. In 1854, Kolbe looked back to Wöhler's experiment and sanctified him as a scientific prophet. "The natural dividing wall that separated organic from inorganic compounds came down," Kolbe declared. Life relied on ordinary chemistry but somehow managed to use it for extraordinary ends.

THIS MUD WAS ACTUALLY ALIVE

On the night of August 14, 1873, Lord George Granville Campbell gazed out from his ship at an ocean on fire. Every wave flashed with light. When Campbell made his way to the stern of HMS *Challenger* and looked down at the keel cutting through the Atlantic, he saw a glowing band of blue and green trailed by rising yellow sparks. When he walked up to the prow, the light coming up from the ocean was bright enough to read by.

It was, Campbell later said, as if the Milky Way "had dropped down on the ocean, and we were sailing through it." But this galaxy, it turned out, was made not of stars but of life.

A sublieutenant in the Royal Navy, Campbell was serving aboard the *Challenger* on a three-year scientific voyage. The ship, originally built for war, had been refitted for research. The navy had installed a hundred miles of ropes, as well as trawls, dredges, and sounding devices. They took out the *Challenger*'s cannons and converted the bays into laboratories. The crew's mission was to learn about the chemistry and biology of the world's oceans. For thousands of years sailors had seen lights on the sea, but now the *Challenger* crew studied the phenomenon scientifically. They first spotted lights near the Cape Verde islands and immediately threw out fine-meshed dredges to see what might be producing them. They hauled up all manner of nocturnal marine creatures, which they brought to the ship's labs to analyze.

As the ship sailed on, it encountered more lights. Sometimes they turned out to be produced by microscopic algae illuminating themselves whenever the water around them was stirred. Sometimes the crew discovered the glow came from siphonophores, monstrous colonies of gelatinous animals stretching as long as sixty feet. Henry Moseley, the ship's naturalist, used his finger to trace out his name on one specimen curled up in a bucket. "The name came out in a few seconds in letters of fire," he said.

The *Challenger* discovered living fire not just at the surface of the ocean but also thousands of feet down. The ship was equipped with new technology to survey the deep ocean, a world almost entirely unknown until then. The crew would periodically drop tethered brass tubes into the depths as the *Challenger*'s engines held the ship motionless against the wind. The tubes fell two miles to the bottom of the ocean, where they measured the temperature—often barely above freezing—and sometimes scooped mud to bring back up. The crew sometimes dragged openmouthed dredges over the seafloor to see what they could haul up. When the dredges were spilled out across the deck, the crew picked through the deep-sea jetsam. Sometimes they found ancient volcanic rocks. Sometimes they found the dust of meteorites that had fallen from space and settled on the seabed. And sometimes they found living things that gave off light: luminescent fish, corals, starfish. The *Challenger* crew wrote long letters about their adventures, which took months to get back to England. But when they arrived, newspapers there and abroad reprinted them. To a Victorian reader, they read like dispatches from an *Apollo* mission to our own planet.

For the *Challenger* crew, some of the most exciting dredges spilled what seemed like nothing but pale mud on the ship's deck. Rather than simply washing it back overboard, they carefully shoveled the mud into filters and preserved what passed through in sealed bottles. In that mud the crew was searching for a primordial creature they called *Bathybius*. Many biologists were convinced that it covered almost the entire ocean floor around the world. It was not an animal or a fungus but a primordial jelly—the same stuff that made up our own cells. On earlier voyages some naturalists had found hints of this mysterious form of life, but the *Challenger* was finally equipped to reveal *Bathybius* in full detail.

No one looked forward to the *Challenger* finding *Bathybius* more than Thomas Huxley, the British scientist who had given it a name.

By the time of the *Challenger* voyage, Huxley had become one of the world's most prominent scientists. He had reached those heights from a childhood mired in filth, poverty, and occasional starvation. Despite his hardships, Huxley's genius still managed to shine. As a child, he taught himself German, mathematics, engineering, and biology. He dreamed of joining an expedition to discover strange new forms of life. A scholarship enabled Huxley to go to medical school, where he quickly proved a master of anatomy. As a teenager, he made a close examination of hair and discovered a hidden sleeve of cells in the sheath that surrounded each strand. It's known today as Huxley's layer.

Staggering debts forced Huxley to leave medical school and enlist in the Royal Navy at age twenty-one as an assistant surgeon. To his delight, he was assigned to HMS *Rattlesnake*, an aging frigate destined for the coasts of Australia and New Guinea, where its crew would search for safe passages. The ship's captain, Owen

Stanley, wanted a doctor with the expertise—or at least the curiosity—to study the animals and plants they'd encounter along the way. "I need not say how gladly I accepted the proffered appointment," Huxley later recalled.

The *Rattlesnake* left England in December 1846. In the South Atlantic, Huxley noticed a Portuguese man-of-war drifting close by, the wind catching the animal's bright blue bladders like a sail. Mindful of its deadly sting, Huxley carefully plucked the man-of-war out of the water and brought it to the ship's chart room. Laying it on the table, he carefully examined its fragile, toxic body until the tropical heat destroyed it. He was dazzled by its anatomy, which was profoundly different from any vertebrate like ourselves. A few naturalists had studied men-of-war before, but Huxley realized they had gotten the anatomy wildly wrong.

As the *Rattlesnake* sailed on toward Australia, he caught more specimens and made a careful study of them. His curiosity extended to other gelatinous creatures such as moon jellies and sea rafts. As he inspected their soft bodies, he found striking similarities between them. They all used the same microscopic harpoons to deliver their stings, for example. All he could do was describe these animals as accurately as he could and send his accounts back to friends in London, hoping they might be read.

When Huxley finally returned to England in 1850 at age twenty-five, his letters had already created a sterling reputation for him. Within a few years he became a professor at the Royal School of Mines and one of the most powerful public champions for science. He wrote essays for magazines and gave lectures intended for "working men." Huxley also found time to keep studying life, working through his collection from his days aboard the *Rattlesnake*. While his expedition days were over, he was now powerful enough to get hold of new samples from Britain's network of shipboard naturalists. Huxley had launched his scientific career skimming the surface of the ocean for strange forms of life, but in the late 1850s his attention plunged down to its dark depths.

A flotilla had begun surveying the ocean floor to prepare for laying the first telegraph cables connecting England to Europe and then to the United States. Like other biologists, Huxley wanted to know whether anything lived down there. He arranged for the surveyors to save some of the mud they brought up, sealing it in jars along with alcohol, in the hopes of preserving any soft tissue that would otherwise rot on the voyage home.

One of the ships that sent Huxley mud was a sounding vessel called HMS *Cyclops*. In June 1857 it left Valentia, Ireland, for Newfoundland, passing along the way over a vast rise of seafloor called Telegraph Plateau. The captain, Joseph Dayman, expected it to be ribs of granite. Instead, his crew hauled up a "kind of soft, mealy substance, which, for want of a better name, I have called ooze."

When the ooze arrived in London, Huxley discovered that it contained odd microscopic buttons. Each button was made of concentric layers that surrounded a central hole. Huxley wasn't sure if they had broken off of animals that lived in the ooze or had fallen from higher in the ocean to their resting place on Telegraph Plateau. Still, they warranted a name, so Huxley dubbed them coccoliths. He filed a short report with the navy and put the ooze on a shelf, where it sat for ten years. It would prove a busy decade for Huxley as he helped usher in a new theory of life.

On Huxley's return home from the *Rattlesnake* expedition in 1850, one of the most important new friends he made was Charles

Darwin. Darwin, forty-one at the time, was known mostly for his own voyage around the world aboard HMS *Beagle*. As far as anyone knew, he had busied himself ever since with barnacles. Darwin and Huxley came from different universes within England: while Huxley had grown up poor, Darwin came from a wealthy family and had never worked for a living. But they immediately recognized that they both shared an obsession with life in all its baffling diversity and were desperate to find a principle to make sense of it all.

In 1856, Darwin invited Huxley to his country house for a weekend visit. There he let Huxley in on a great secret: like his grandfather Erasmus, Charles Darwin had become convinced that life evolved. But Charles had not produced a poem about the idea. Instead, he had developed a detailed theory, which he explained to Huxley. Natural selection turned old species into new ones, into new forms of life. Every species, Darwin argued, was merely a branch on the tree of life.

Before that moment Huxley had been skeptical about evolution, but now he began to warm to it, recognizing that Darwin had succeeded where others had failed. While Darwin holed up on his country estate, Huxley championed his theory on the lecture circuit and in magazines. He called on his fellow biologists to carry Darwin's project further, to join all the branches of the tree of life together. As they came to better understand the evolutionary tree, they would be able to make their way down to the base—to the stage of history when life first arose. "If the hypothesis of evolution is true, living matter must have arisen from not-living matter," Huxley declared.

The best place to search for evidence of that transition, Huxley decided, was the ooze.

Huxley's suspicions had a long pedigree. Their origins reached back over a century, to Abraham Trembley's work with polyps.

Trembley had discovered a jellylike substance in the animals that seemed endowed with vital force. Albrecht von Haller recognized this substance in the animals he dissected and speculated that it was responsible for irritability. The vitalists who followed Trembley and Haller went even further. They claimed that the goo was the stuff of life, found in every species.

A German biologist named Lorenz Oken even gave this gelatinous mass a name: *Urschleim*, or primal slime. Oken envisioned the primal slime as a vast, continuous substance that formed spontaneously on the early earth. It then broke up into microscopic blobs of living matter, which then evolved into complex life as we know it. Yet even today, Oken argued, the primal slime continued to go through cycles of creation and destruction inside all living things.

Oken championed *Urschleim* in wildly Romantic speculations that had no basis in experimental evidence. Nevertheless, even more sober-minded biologists gradually came to agree that life was built from a universal goo. In the 1830s a French zoologist named Félix Dujardin found a "living jelly" inside single-celled microbes. More evidence came from microscopic studies on the tissues of plants and animals, which revealed that they were masses of cells. When nineteenth-century biologists looked inside cells, they always found the same living jelly. "The cell was redefined around a frothing lump of mucus," writes the historian Daniel Liu.

This mucus moved and quivered. It pushed cells from within. "I dare not venture to express the slightest suspicion of the cause of this motion," the German biologist Hugo von Mohl declared in 1846. Within a few years scientists agreed to call this mysterious frothing mucus "protoplasm." And soon a suspicion emerged that protoplasm didn't just have a vital power of motion; it might also carry out the chemistry that produced organic molecules. It might organize the interiors of cells. It might pull apart one cell to make two and drive the development of cells into complex embryos. It seemed like there was nothing protoplasm couldn't do.

Huxley, while not a cell biologist or chemist himself, kept close watch on the growing evidence for protoplasm as the basis of life. If evolution was like a river flowing through time, he recognized, protoplasm was its water. It was protoplasm that was passed down from one generation to the next, and somehow generating evolution's new forms. "If all living beings have been evolved from pre-existing forms of life," Huxley wrote, "it is enough that a single particle of living protoplasm should once have appeared on the globe."

In the early 1860s researchers in Canada discovered what looked like fossil protoplasm. From some of the oldest rocks known to scientists at the time, they found fossils of a speck-sized, shell-covered creature. The biologist William Carpenter, who examined the organism carefully under a microscope, described it as "a little particle of apparently homogeneous jelly."

Carpenter called the new species *Eozoön,* or dawn animal. When Darwin read about it, he updated his 1866 edition of *The Origin of Species* to include the discovery as further evidence for evolution. "After reading Dr. Carpenter's description of this remarkable fossil, it is impossible to feel any doubt regarding its organic nature," he declared.

Geologists found more *Eozoön*, uncovering vast sheets of fossils in Canada and beyond. Judging from the different layers where they found the fossils, *Eozoön* seemed to have endured for vast stretches of time. In fact, Carpenter declared at a geology meeting in London that he "should not be astonished even if such a structure as *Eozoon* were found in deep-sea dredgings of the present day."

In 1868, shortly after Carpenter published his study of *Eozoön*, Huxley did something odd: after a decade, he took the ooze from the *Cyclops* off its shelf to give it a fresh look. No one knows exactly why he decided to break the ten-year spell. Maybe he thought *Eozoön* was still alive on the bottom of the ocean. Maybe he thought the ooze contained the primal slime predicted by Oken. Maybe he was just excited to try out the powerful new microscopes he had just acquired.

Whatever the reason, Huxley looked at his ooze, and the ooze now gave Huxley a start. He saw something in it that had not been visible before: "lumps of a transparent, gelatinous substance." The substance formed a blobby network across Huxley's field of view, scattered with the tiny coccolith buttons as well as strange "granule-heaps," as he called them.

If Huxley looked long enough, the lumps moved. He concluded that this gelatinous substance was protoplasm. He must be looking at "simple, animated beings." If the ooze collected by the *Cyclops* was typical of the Atlantic, then the whole ocean might be covered by what he called a "deep-sea 'Urschleim.'"

Huxley concluded that in this slime he had discovered a species in its own right, unlike any form of life previously found, which he named *Bathybius haeckelii*. *Bathybius* meant deep life, and *haeckelii* honored the German biologist Ernst Haeckel, the leading proponent that all life evolved from a simple, protoplasm-filled ancestor. "I hope that you will not be ashamed of your godchild," Huxley told Haeckel.

Huxley unveiled *Bathybius* in August 1868 at a scientific meeting. A reporter there marveled at the idea of a "living paste on the floor of the Atlantic." Huxley presented *Bathybius* as evidence of a sweeping theory about life—about its very nature and its entire history. In the months that followed, he roamed Britain giving a series of lectures on the physical basis of life. From one city to the next, he made a profound impression in the crowded halls and churches where he spoke. "The audience seemed almost to cease to breathe, so perfect was the stillness," a journalist reported from his talk in Edinburgh.

"What hidden bond can connect the flower which a girl wears in her hair and the blood which courses through her youthful veins?" Huxley asked his listeners. The answer was protoplasm. "It may be truly said that the acts of all living things are fundamentally one," Huxley said.

Protoplasm was simply an arrangement of organic molecules whose functions no one yet understood, Huxley declared, but which ordinary physics would suffice someday to explain. There was no need to imagine a mysterious vitality in living things. That made as much sense as saying water had "aquosity."

Ministers might tell their congregation that everything comes from dust and to dust returns. But protoplasm revealed a different cycle, one in which life turns to life. "I might sup upon lobster, and the matter of life of the crustacean would undergo the same wonderful metamorphosis into humanity," Huxley said. "And were I to return to my own place by sea, and undergo shipwreck, the crustacea might, and probably would, return the compliment, and demonstrate our common nature by turning my protoplasm into living lobster."

The delicate balance of scandal and science Huxley delivered proved a smash. Three months after his lecture in Edinburgh, the text was published in *Fortnightly Review* as an essay called "On the Physical Basis of Life." Now protoplasm became famous far beyond Scotland. The issue of *Fortnightly Review* went into seven editions to meet the demand, and newspapers abroad reprinted great swaths of it.

As Huxley scurried from lecture to lecture around England, a scientist named Charles Wyville Thomson was sailing a small steamship called the HMS *Lightning* north of Scotland. By the 1860s, scientists like Thomson wanted to study the ocean for the ocean's sake. He wondered just how much life existed in the deep ocean: Was it an underwater desert or a jungle? The Admiralty provided him with the *Lightning*, a small converted gunboat, for a trial run. Thomson and his crew scooped up bits of the seafloor, and sometimes they brought up an oddly sticky hunk of mud. Mindful of the newly discovered *Bathybius*, they looked at the mud under a microscope and saw movement. It had a strange egg-white appearance, like protoplasm.

"This mud was actually alive," Thomson declared.

After the six-week voyage of the *Lightning*, Thomson delivered the mud to Huxley, who pronounced it a second sample of *Bathybius*. Still more *Bathybius* came to light in the South Atlantic and the Pacific. In August 1872 American explorers searching for the North Pole found what looked like an even more primitive version of *Bathybius* in the Arctic Ocean, which they dubbed *Protobathybius*.

With his primitive creatures turning up all over the world, Huxley now saw *Bathybius* as a kind of global carpet. "It probably forms one continuous scum of living matter girding the whole surface of the earth," he said.

Some scientists rejected all this evidence and denied that *Bathybius* existed. A biologist named Lionel Smith Beale called it "fanciful and improbable." But Beale did not attack Huxley out of some disinterested skepticism. He was a vitalist, and he saw *Bathybius* as a threat to the fundamental divide between life and everything else. "Life is a power, force, or property of a special and peculiar kind, temporarily influencing matter and its ordinary forces, but entirely different from, and in no way correlated with, any of these," Beale wrote.

For the most part, though, scientists saw the discoveries of *Bathybius* around the world as proof that it was real. In 1876 a zoology textbook put *Bathybius* and its shredded tapestry of protoplasm on its first page. In Germany, Haeckel was as delighted by Huxley's discovery as the critics were appalled. *Urschleim*, he said, "has become a complete reality through Huxley's discovery of *Bathybius*." Haeckel shared Huxley's new vision of the planet, declaring that "huge masses of naked, living protoplasm cover the greater ocean depths."

Haeckel wondered where those huge masses came from. "Is protoplasm perhaps originating continually through spontaneous generation?" he asked. "Here we stand before a series of dark questions, the answers to which can only be hoped for from subsequent researches."

Charles Wyville Thomson used his success aboard the *Lightning* to win support for a survey of the deep sea across the entire globe. When the *Challenger* expedition came together, Thomson was appointed its scientific director. The ship had a captain, but it was Thomson who was really in charge. He oversaw a staggering amount of research into biology, geology, and meteorology. The *Challenger* crew collected birds of paradise and seaweed and human remains. They prepared reports on the plants of Bermuda, on the chemical composition of the oceans, on barnacles. Ultimately their data would fill fifty volumes. Thomson would be long dead by the time the last volume came out.

But amidst all their shipboard work, the crew of the *Challenger* always made time to search for *Bathybius*. They had every reason to expect to find it in abundance, and they were eager to study the fresh samples in their shipboard laboratory, rather than just store them away for the long voyage home.

It took a few weeks for the crew to get adept at scooping up deep mud on their way across the Atlantic. John Murray, Thomson's second-in-command, began carefully skimming off water from the mud's surface, where he believed fresh *Bathybius* would most likely be found. He put the samples under the ship's powerful microscopes and searched them for hours, looking for the blobby networks of protoplasm that so many others had found.

He found nothing.

As each sample came up, Murray and his colleagues would store some of the mud in jars with alcohol so that Huxley and other scientists back home might someday study it and perhaps have better luck. One day Murray glanced at the jars and observed that a translucent layer had formed on top of some of their mud. He took the jars down and inspected the layer. It had the consistency of jelly.

The ship's chemist, a wealthy young Scotsman named John Buchanan, was intrigued by Murray's discovery. Perhaps what previous scientists had taken for *Bathybius* was not some form of life in the sea floor ooze, but a jelly-like byproduct of chemical reactions that took place in the jars. To test these possibilities, he let a sample of deep-sea water evaporate. "If the jelly-like organism which had been seen by some eminent naturalists in specimens of ocean-bottom and called *Bathybius* really formed, as was believed, an all-pervading organic covering of the sea-bottom, it could hardly fail to show itself when the bottom-water was evaporated to dryness and the residue heated," he later wrote.

But it did fail. Once the water evaporated, Buchanan could find no organic remains.

He turned to the jelly that Murray had spotted in the jars. His experiments revealed that they contained no organic matter either. Buchanan found instead calcium and sulphate—gypsum, in other words. As the *Challenger* sailed from Hong Kong to Yokohama, Buchanan ran more experiments and realized what had happened. Putting the deep-sea mud in alcohol had driven the calcium and sulphate to form a jellylike mass.

With a few shipboard experiments, Buchanan and Murray had wiped the planet clean of its most primordial, fundamental form of life. They wrote up their obituary for *Bathybius* in cold, clinical prose. "In placing it amongst living things," Buchanan concluded, "the describers have committed an error."

You might expect that Thomson responded by quelling his team's blasphemy. After all, seven years earlier, on the other side of the world, Thomson himself had dredged up *Bathybius*. He had written glowingly about the species in a best-selling book about the sea before setting sail on the *Challenger*. But Thomson held on to his convictions lightly. Buchanan and Murray convinced him their science was good, and so, on June 9, 1875, Thomson composed a letter to Huxley to relay the bad news.

"You should be told exactly how it stands," he told Huxley. "None of us have ever been able to see a trace of *Bathybius*, although it has been looked for throughout with the most utmost care." Murray and the other members of the team, Thomson wrote, "deny that such a thing exists."

When Huxley received the letter, he did not hide the disastrous message. Instead, he passed it on to the journal *Nature* for publication, with a note of his own at the end: "I am mainly responsible for the mistake, if it be one."

By the time the *Challenger* returned to England on May 24, 1876, *Bathybius* was pretty much dead. Among its few remaining defenders was Haeckel, who was dismayed to see Huxley give up the fight for his namesake. "The more the real parent of *Bathybius* shows himself inclined to give up his child as hopeless, the more I feel bound, as its godfather, to look after its rights," he once said. But Haeckel had nothing to offer against the *Challenger*'s evidence. *Bathybius* soon disappeared from textbooks, dismissed as a spectacular mistake. Its fossil forerunner, *Eozoön*, soon followed the same path to obscurity—a deceptive crystallization, it turned out, rather than the mark of ancient primitive life.

In fact, it was Huxley's enemies who did the most to keep the memory of *Bathybius* alive. In 1887 the Duke of Argyll, the leading opponent of Darwinism in the nineteenth century, revived the embarrassment to question Huxley's entire view of life. The duke called the affair "a case in which a ridiculous error and a ridiculous credulity were the direct result of theoretical preconceptions. *Bathybius* was accepted because it was in harmony with Darwin's speculations."

Huxley, who had a low opinion of the Duke of Argyll for not doing any science of his own, freely granted that he had made an error. But, he added, "the only people, scientific or other, who never make mistakes are those who do nothing."

Still, the duke did have a point, as the historian Philip Rehbock later observed. "*Bathybius* was a highly functional concept," Rehbock wrote, "an explanatory device which made sense in the context of mid-nineteenth-century biological and geological thinking."

In the borderland between the living and the nonliving, conceptual mirages have a way of taking shape and gaining fame. Yet, despite his planet-wide error, Huxley's reputation remained intact. When he died in 1895, the *Proceedings of the Royal Society of London* sang his praises in an obituary that ran twenty pages.

"Whatever bit of life he touched in his search, protozoan, polyp, mollusc, crustacean, fish, reptile, beast, and man—and there were few living things he did not touch—he shed light on it, and left his mark," the journal declared. In all those pages, they couldn't find room to mention *Bathybius.*

A generation later, John Butler Burke would suffer a harsher fate when his radiobes proved false. Although Huxley had been fooled by a mirage, he was still right about life's big picture. Evolution is real, and protoplasm does indeed unite all of life. But it does so with a bond far more intricate than Huxley could have imagined.

A PLAY OF WATER

Bathybius was dead, but protoplasm lived on. As the nineteenth century drew to a close, its inner workings slowly began to emerge. Some of the first clues came not from the seafloor but from beer.

Throughout history, making beer had been a kind of alchemy. People began brewing at least 13,000 years ago, when glaciers covered New York and woolly mammoths trundled across Siberia. The first brewers, who lived somewhere in the Near East, collected wheat and barley plants and boiled them into a concentration of sugar known as a wort. They then waited for the wort to ferment into a bubbling brew that could get them drunk. What happened during fermentation was anyone's guess.

In the nineteenth century, chemists offered one answer and microbiologists offered another. The chemists, working in the tradition of Friedrich Wöhler, thought of fermentation in terms of molecules becoming new compounds. To them, it seemed that plant sugar underwent chemical reactions that produced alcohol and other molecules, along with bubbles of carbon dioxide gas.

The microbiologists meanwhile looked at fermentation as an act of life. The dregs in the wort—long known as yeast—turned out to be composed of living single-celled organisms. Fermentation could not take place without them. For thousands of years, brewers had unknowingly inoculated their beer by leaving their wort open to the air. Drifting yeast spores naturally settled on it and took over the fermentation from there. Life was essential to the process. A sterilized wort could never become beer. By the end of the nineteenth century, microbiologists had turned brewing into a kind of industrial biology. Brewers could pick out which species of yeast they wanted to use, ensuring that their beer would wind up with a predictable flavor. Every pint of beer raised in every pub seemed proof of life's vital force. When sugar came into contact with living matter, it turned to alcohol in a reaction that could otherwise never take place.

The chemists were not impressed. The notion that tiny cells of yeast guzzled wheat and magically pissed out alcohol seemed like absurd vitalism.

A young German chemist named Eduard Buchner won a Nobel Prize by trying to broker a truce between the two sides of the great beer debate. By the late 1800s, scientists knew that living things made a special class of proteins called enzymes that were remarkably good at breaking down certain other molecules. Some researchers proposed that yeast contained an enzyme that could cut apart sugar. Yes, yeast were essential to fermentation, but, no, they contained no vital forces.

In the 1890s Buchner set out to find these imaginary enzymes. He mixed yeast powder with a fine grit and then ground it in a mortar into a dark, damp dough. The membranes of the yeast cells tore open, dumping out their protoplasm.

Buchner spread this new concoction on a flat surface and crushed it under a hydraulic press. Out came a pleasant-smelling yeast juice. To kill any cells that managed to slip into the juice, Buchner added arsenic and other poisons. The juice was now completely robbed of life.

And yet, when Buchner added sugar to this lifeless juice, it gave off a fizz of carbon dioxide bubbles and turned to alcohol. Fermentation did not depend on living cells, Buchner's experiment showed. It didn't even need bits of living protoplasm. An ordinary enzyme must be responsible.

At first the idea seemed outrageous to biologists and brewers alike. They could not conceive of protoplasm as a jumble of specialized molecules, each carrying out its assigned reaction. One expert on fermentation predicted that Buchner's claim "will enjoy none too long a life."

But soon other scientists succeeded in repeating Buchner's experiment, and they pushed the work even further, isolating Buchner's enzyme and giving it the name *zymase*. The French microbiologist Émile Duclaux declared that Buchner was "opening up a new world." It was a world of biochemistry, in which living things were filled with a zoo of active proteins.

When Buchner went to Stockholm in 1907 to accept his Nobel Prize, he tried to play the part of peacemaker. Mechanists and vitalists didn't have to fight over fermentation. The vitalists had been right that yeast is essential to fermentation. Enzymes couldn't exist without this living thing to create them. But yeast did not use mysterious vital forces to ferment beer. They made zymases: ordinary molecules that followed the ordinary laws of chemistry. Removed from a cell, an enzyme was lifeless—but it could still carry out the same chemical reactions.

"The differences between the vitalistic view and the enzyme theory have been reconciled," he announced. "Nobody is ultimately the loser."

If Buchner imagined he could broker a truce in a war that had lasted by then for over two centuries, he must have been deeply disappointed. The arguments over the nature of life only grew louder in the years after he picked up his prize. The biochemical vision of life—like the mechanical ones that came before—left many scientists dissatisfied. It was all well and good to find one enzyme that broke down sugar and another that broke down starch. But no one could piece together a few such reactions into the grand transformations essential to life—the way plants turned sunlight into roots and flowers, for example, or how a single cell became a human being. As microscopes grew even more powerful, biologists were discovering that protoplasm was actually as busy as a city, crammed with compartments, filaments, and granules. No one could tell yet what went on in those secret chambers or how many were even real. Some appeared one day under the microscope and then vanished the next.

"Which of them are alive? Which of them, if any, constitute the physical basis of life?" asked the American cell biologist Edmund Wilson in 1923. "These are embarrassing questions."

Some scientists argued that these questions would remain embarrassing forever. The simple chemistry carried out by enzymes could not guide an egg into an embryo. Trembley's polyps needed more than molecules to rebuild their bisected bodies. But the scientists who rejected a purely mechanistic view of life weren't arguing for a mystical vital force, either. What made life special was that it existed on more than one level.

Lower levels spontaneously gave rise to higher ones. One enzyme might only be able to do one thing—fuse two molecules together, for example—but bring together billions of enzymes, carrying out billions of different tasks, and you suddenly had a cell. Step up to the next level, and a group of cells became a body. Bodies combined into populations, populations into ecosystems.

Once you hopped to a new level, you had to stay there to make sense of it. Try to understand a cell by breaking it back down to enzymes, and you kill it. The cells inside the bodies of snowshoe hares cannot explain the booms and busts their populations go through across Canada every few years. The answer lies in the bloody dance of hares and lynx.

The public at large followed these debates closely. The emerging science of biochemistry seemed poised to give humanity a Frankensteinian power over life. But in the process, it seemed as if it would reduce life—especially human life—to depressingly small bits. Memories, emotions—our very selves—seemed to dwindle down to blind jostling of proteins. People wanted more from their lives, more from life itself, and vitalism seemed to offer what they craved: a vital force that lay beyond the biochemist's reach.

At the dawn of the twentieth century, the vital force grew into something like a religious phenomenon: the human spirit to some, the spark of the divine to others. The French philosopher Henri Bergson gained a huge following with his claims that all of life shared a vital impulse, or *élan vital.* "Life is, more than anything else, a tendency to act on inert matter," he wrote in his 1911 book, *Creative Evolution.* Murky and meandering, it nevertheless became a best seller. When Bergson traveled to New York to deliver a series of lectures, he reportedly created the city's first traffic jam. A thousand people showed up just to gawk at him as he sipped tea with the wives of Columbia University professors.

Bergson and the other neo-vitalists did not impress the biochemists. In a 1925 essay, the British scientist Joseph Needham declared that they "have won no confidence at all among the research workers in biochemistry and physiology." Talking of vital forces was nothing but a celebration of ignorance. In the nineteenth century, many physicists sought to explain how light traveled through space by claiming it was filled with a substance called ether. It was weightless, transparent, frictionless, and undetectable—and yet was supposed to suffuse the universe. As soon as modern physics emerged, it proved to be a fiction. In the early 1900s, biochemists like Needham felt confident that vital forces also disappear, remembered as life's ether.

Needham accepted that all of life could not be explained merely in terms of atoms. It had many levels, each of which deserved attention. But that was no reason to abandon a mechanistic foundation. Even if a single enzyme could not explain an eagle, it was certainly a good place to start. In the 1920s, biochemists were discovering how enzymes worked together in teams. One enzyme might cut part of a molecule off and then hand over the molecule to another enzyme to change it in another way. Gradually these chains of enzymes grew into great interlocking loops of metabolism. And, meanwhile, what were the vitalists discovering? Over and over again they did nothing but point to the open questions scientists had yet to answer. To Needham they were no better than nineteenth-century theologians, denying evolution by pointing to gaps in the fossil record.

"In the laboratory," Needham sighed, "it simply will not do."

Needham's words proved prophetic. As the twentieth century rolled on, vitalism yielded more of its ground to chemistry and physics. Even irritability, that fundamental force that seemed unique to life, yielded to the research of an extraordinary Hungarian physiologist named Albert Szent-Györgyi. Before he was done, he could conjure irritability on demand.

"My inner story is exceedingly simple, if not indeed dull," Szent-Györgyi said late in life. His existence was devoted to science, full stop. As for his external life, Szent-Györgyi acknowledged that it had been "rather bumpy." That's putting it mildly. Some of those bumps would have gotten most people killed.

Szent-Györgyi was a medical student when World War I broke out. He joined the Hungarian army and served for three years, until he could see that the war was lost and fighting any longer was a senseless sacrifice. "The best service I could do for my country was to stay alive," Szent-Györgyi wrote. "So, one day, when in the field, I took my gun and shot myself through the bone of my arm."

His self-inflicted wound allowed Szent-Györgyi to return to Hungary just in time for a Communist uprising. His family lost virtually all their possessions, and he fled the country with his wife and child. In Prague and then Berlin, they verged at times on starvation. Szent-Györgyi managed to continue his medical studies, but in time he came to realize that he didn't actually want to cure people. "I wanted to understand life," Szent-Györgyi said.

To do so, he joined the efforts to dissect protoplasm. He studied how enzymes cooperated inside our cells in the transformation of food to fuel, eventually earning a PhD at the University of Cambridge. The reactions Szent-Györgyi uncovered would turn out to be key steps in the loops of metabolism that keep us alive. In enzymes Szent-Györgyi saw a unity in life. "There is no basic difference between man and the grass he mows," he said.

He proved it with a discovery that earned him the Nobel Prize, a discovery that started with his puzzling over potatoes and lemons. When potatoes are cut, they turn brown, but lemons do not. Szent-Györgyi reasoned that oxygen reacted with a compound in the potatoes, but the lemons contained a second compound that slowed those reactions down.

He searched for that second compound for years, eventually finding it in the cells of many plants as well as some animals.

When Szent-Györgyi was ready to publish a paper on the molecule in 1928, there was still much about it he didn't understand. If you had asked him about it, he would have shrugged and said, "God knows." In fact, he asked his editors at the *Biochemical Journal* if he could name the molecule "Godnose," just to make his ignorance clear. They forced him to call it hexuronic acid.

It later came to be known as vitamin C. Scientists determined that it's essential for repairing cellular damage, building proteins, and many other functions. While lemons and some other plants have genes for making vitamin C, we humans have to get our supply in our food. Szent-Györgyi's discovery made it possible to synthesize the molecule from scratch, but he refused to put a patent on it, believing vitamin C belonged to all humanity. It didn't make him rich, but he did get a summons from Stockholm.

At age forty-four, with a Nobel Prize, Szent-Györgyi finally considered himself ready for serious science. "I felt I had now enough experience for attacking some more complex biological process, which could lead me closer to the understanding of life," he said. He chose to study muscle. "Its function is motion," Szent-Györgyi said, "which has always been looked upon by man as the criterion of life."

At the University of Szeged in Hungary, where Szent-Györgyi was appointed a professor, he put together a team of young scientists to take on the mystery that had vexed Albrecht von Haller two centuries earlier: how muscles moved. He knew that if you soaked muscles in a solution of salt, their cells released a viscous ooze. In the ooze were filament-shaped proteins called myosin that many scientists suspected generated the force that made muscles contract.

Another molecule that caught Szent-Györgyi's fancy was ATP. It had been discovered in 1929, but no one knew yet what it was for. Some researchers suspected that muscles used ATP as fuel, capturing the energy released when its bond was broken. In 1939, Szent-Györgyi learned that Russian biologists had discovered myosin could grab ATP molecules and split them. Szent-Györgyi decided to look more closely at that reaction.

As he began this new line of research in the late 1930s, Szent-Györgyi became cut off from the world. Hungary had formed a loose alliance with Nazi Germany against Russia in the hopes of getting back some of the land they lost in the Treaty of Versailles. Britain then declared war on Hungary, and the country became isolated behind the Axis lines. Over the course of his career, Szent-Györgyi had built up an international network of collaborators. Now he and his colleagues at Szeged had to work alone.

Soon his lonely team of scientists saw something extraordinary. They isolated threads of myosin and dropped them into boiled muscle juice. In a matter of seconds, the long, translucent threads scrunched into dark stubs. Szent-Györgyi and his colleagues were seeing muscles contract on a molecular scale.

To understand how this motion occurred, the researchers stripped the boiled muscle juice to its bare essentials. They prepared a solution containing just ATP, along with some potassium and magnesium to keep the cells working properly. Those three ingredients were enough. When the scientists dropped myosin threads into this mixture, the proteins contracted. They had recreated one of life's most basic functions in a test tube.

One member of Szent-Györgyi's team, Bruno Straub, called it "the most beautiful experiment I ever witnessed." Another, Wilfried Mommaerts, said it was "possibly the greatest biological observation." Part of what made it so great, Mommaerts felt, was its simplicity—"the hallmark of true genius."

This work of genius was all the more remarkable because Szent-

Györgyi was splitting his time between science and spy craft. He had been appalled by Hitler's rise and helped Jewish scientists escape from Germany. At Szeged he stood down mobs of fascist students hunting for Jews at the university. When Szent-Györgyi received his Nobel Prize winnings, he invested them only in stocks that wouldn't benefit from the war economy. (He lost it all.) And once the war broke out, Szent-Györgyi quietly joined a resistance group.

In 1943 he boarded a train for Istanbul on a secret mission. For his cover, he delivered a scientific lecture at a Turkish university. But he then met secretly with British intelligence agents, letting them know that Hungary might consider switching sides and joining the Allies.

Returning to Hungary, Szent-Györgyi believed his mission was a success. He was wrong. Nazi spies learned of his betrayal, and Hitler screamed for his extradition to Germany. The Hungarian government tried to placate Hitler by putting Szent-Györgyi under house arrest. He managed to slip away and remained in hiding for months, staying one step ahead of the Gestapo as they massacred fellow members of the resistance. All the while Szent-Györgyi's team at the university continued running their experiments on muscles and writing up their results. From time to time Szent-Györgyi would unexpectedly turn up at the lab in Szeged to check on their progress and then vanish again.

Staying alive was less important to Szent-Györgyi than letting the world know about his experiments. If the Gestapo put a bullet in his head, the world might never know what he and his colleagues had done. Szent-Györgyi arranged for a few hundred copies of their papers to be printed up, but he struggled to get them to friends outside of Hungary. Eventually Szent-Györgyi found a place where he thought he could hide safely: the Swedish legation in Budapest. But his cover was blown when a Swedish scientist sent a wire to the legation letting Szent-Györgyi know he had gotten the manuscript about his work on muscles.

The Gestapo prepared to storm the legation, eager to catch the spy who had eluded them now for months. When the legation learned of their impending attack, the Swedish ambassador drove off in a limousine with Szent-Györgyi hidden in the trunk.

The war came at last to Hungary. Nazi and Soviet forces began fighting for Budapest, destroying it in the process. Szent-Györgyi hid in bombed-out buildings in the no-man's-land between the two armies until the Soviet foreign minister dispatched a squadron to find him. They whisked him and his family to a Soviet military base south of Budapest where they lived for three months, until the war was over and they could return home.

Szent-Györgyi returned to the ruins of Budapest a national hero. And the scientific world—which had feared he was dead—marveled at the 116-page report published in the journal *Acta Physiologica* in which he and his colleagues explained their solution to one of life's mysteries.

Szent-Györgyi thought at first that the Soviet Union would help Hungary become a thriving postwar democracy. He set to work rebuilding his homeland's scientific establishment, and gossip spread that he might soon be elected president. It didn't take long for Szent-Györgyi to recognize that Hungary had traded an old oppressor for a new one. The Soviets began torturing dissidents, then killing them. Szent-Györgyi reached out to contacts in the United States, hoping to land a job as a professor at an American university. But the American government saw his friendly dealings with his Soviet overlords as a sign he might be a spy rather than a Nobel-winning refugee.

As part of his campaign for entry, Szent-Györgyi traveled to

Boston to give a series of lectures at MIT. There he told his American audience the story of his wartime work on muscles. He spoke of filaments and myosin, of ATP and ions. And once he made his way through his findings, Szent-Györgyi stopped to reflect on what he had learned.

"I have come to my journey's end and now you probably expect me to finish my lecture in a dramatic way by telling you what life is," he said.

Biochemists had been doing this for decades now. In 1911, the Czech scientist Friedrich Czapek crafted a succinct definition: "On the whole what we call life is nothing else but a complex of innumerable chemical reactions in the living substance which we call protoplasm."

Over the course of his own career, Szent-Györgyi came up with definitions of his own, if only to mock the idea that a simple definition would ever be possible. "Life," he liked to say, "is just the play of water."

Plants and bacteria split water through photosynthesis to build carbohydrates. And in the respiration of cells—either in the plants or in animals like us that eat the plants—the liberation of the energy in those carbohydrates requires putting the water molecules back together again. "What we call 'life' is a certain quality, the sum of certain reactions of systems of matter, as the smile is the quality or reaction of the lips," Szent-Györgyi once said.

When he stopped to reflect more deeply on what he and his fellow biochemists were learning about life, Szent-Györgyi found it hard to offer a meaningful definition. If the definition of life involved something that sustained itself through chemical reactions, then a candle flame might be alive. What about a star, or a civilization?

All living things, Szent-Györgyi explained to his audience at MIT, shared some hallmarks. But thinking too categorically about those hallmarks was a one-way ticket to absurdity. “One rabbit could never reproduce itself,” Szent-Györgyi observed. “And if life is characterized by self-reproduction, one rabbit could not be called alive at all.”

We can find different features of life at different scales, Szent-Györgyi said, but only depending on the features of life we cherish most. “The noun ‘life’ has no sense,” Szent-Györgyi declared, “there being no such thing.”

Soon after his visit to MIT, Szent-Györgyi won permission to move to the United States. But his attempts to land a professorship failed, and he wound up on Cape Cod in Massachusetts with a tenuous connection to the Marine Biological Laboratory. Still, he made the most of life in his new homeland. Each summer he hosted scientists at his rambling seaside house in the village of Woods Hole. He grew famous for his parties, for his nighttime expeditions to fish for striped bass, for leading armadas of backstrokers around a nearby peninsula, for dressing up for parties as Father Time or Uncle Sam or Saint George, armed with an aluminum foil sword and shield.

Szent-Györgyi also kept doing research in Woods Hole, supporting his efforts by creating an institute funded by patrons. There he opened a new line of research to find the fundamental difference between living and nonliving matter.

Living things were endowed with a special sort of chemistry, which Szent-Györgyi called “subtle reactivity and flexibility.” He believed life gained this power from the electrons that shuttle from atom to atom within proteins. Molecules like vitamin C, he believed, could move electrons from oxygen to other molecules without causing damage inside a cell. “It is involved in bringing matter to life,” Szent-Györgyi declared.

His intuition pointed him in the right direction. To stay alive, cells must manage their electric charge and prevent charged compounds from pinging around their interiors, destroying DNA and proteins. But Szent-Györgyi, with no training in quantum physics, had at last gotten in over his head. Ever the showman, he confidently promised that he would learn how matter is brought to life—and thereby find a cure for cancer.

Not long before Szent-Györgyi died in 1986, he made an extravagant request to the National Institutes of Health for millions of dollars. John Edsall, a Harvard biologist and a longtime admirer of Szent-Györgyi, reviewed his application and visited his lab to take stock of his work. Finding little in Woods Hole to inspire confidence, Edsall turned him down.

"I felt, with pain, that he had lost the special touch and instinct that had guided him aright in his brilliant pursuit of significant problems in the past," Edsall said. Nothing could take away Szent-Györgyi's Nobel Prize or his wartime discoveries about muscles. But his colleagues were sad to see the mystery of life finally have its vengeance on him. What made it all the sadder was that Szent-Györgyi could see what was happening as well, as documented in an essay he wrote in 1972:

"I moved from anatomy to the study of tissues, then to electron microscopy and chemistry, and finally to quantum mechanics. This downward journey through the scale of dimensions has its irony, for in my search for the secret of life, I ended up with atoms and electrons, which have no life at all. Somewhere along the line, life has run through my fingers."

SCRIPTS

In the 1920s the world was still coming to terms with the weirdness of quantum physics. People could be forgiven for thinking that physicists had lost their minds. Up until then they had presided over a stately, predictable cosmos that followed Newton's clocklike laws, and now they were announcing that the foundations of that cosmos defied common sense. Light was both a particle and a wave. An electron could be here and there at once. Energy was a series of quantum jumps.

But when Max Delbrück discovered this new world as a physics student in Germany, he immediately felt at home. He impressed his teachers with his ability to uncover new implications of the theory of quantum physics and use them to explain properties of real atoms. Delbrück might have gone on to a successful career doing just that if he hadn't traveled to Denmark in 1931. He went there to study under the Nobel Prize–winning physicist Niels Bohr, only to discover that Bohr didn't consider quantum physics the strangest thing in the world. Life was stranger.

Niels Bohr argued that physicists would never be able to see all physical reality at once. If they wanted to study light, for example, they could study it as a particle or as a wave but not both at the same time. Bohr believed life had a two-sided nature as well. A physicist could make sense of the gases and liquids in a body, but physics could not explain how a body could keep its gases and liquids stable in order to survive.

"He talked about that a lot," Delbrück later recalled of Bohr. "You could look at a living organism either as a living organism or as a jumble of molecules."

Bohr helped Delbrück to see life as a frontier where a physicist might be able to discover something radically new. "If one looks at even the simplest kind of cell, one knows it consists of the usual elements of organic chemistry and otherwise obeys the laws of physics," Delbrück said. "One can analyze any number of compounds in it but one will never get a living bacterium out of it, unless one introduces totally new and complementary points of view."

Life maintains an extraordinary kind of order, even though the universe seems purpose-built to tear order apart. It's not surprising to see a wineglass fall to the ground and shatter into a hundred shards. It is surprising to see a hundred shards assemble into a wineglass. Heat a pot of water, squirt an assortment of food dyes into it, and you don't expect to see them organize into a beautiful rainbow. You see the color of mud. Life defies this directive. Eggs hatch into swans, and seeds sprout into zinnias. Even a single cell can maintain an astonishing molecular order.

"The meanest living cell becomes a magic puzzle box," Delbrück later explained, "full of elaborate and changing molecules, and far outstrips all chemical laboratories of man in the skill of organic synthesis performed with ease, expedition, and good judgment of balance."

After he finished studying with Bohr in Denmark, Delbrück returned to Germany, where he worked in the Berlin laboratory of the physicist Lise Meitner. By day he worked on questions such as how to steer the path of gamma rays. At night he tried to learn biology pretty much from scratch. Delbrück felt as if he were the only person on Earth who had taken up Bohr's mission.

"I mean the physicists didn't know enough biology, and didn't care about it on the whole," Delbrück said, "and the biologists, for them anything like quantum mechanics was utterly beyond their ken."

Eventually Delbrück found a few other people wandering these borderlands: "a group of, as it were, exiled, internal exiled, theoretical physicists," he called them. Delbrück had admired the way Bohr built a little society of physicists in Copenhagen to explore quantum physics together, and so he followed suit in Berlin. He invited his new friends to gather for meetings at his mother's house. The agenda of the gatherings was "to jointly consider some of the riddles of life."

Whenever Buchner ground yeast, out came zymase. But when biochemists pulled apart the cells of other species, they found other enzymes instead. How did zymase wind up in yeast cells and not in our own? And why was it that when one yeast cell divided in two, the new cells still maintained their own supply of zymase? When Delbrück turned to biology in 1932, biologists had only hazy guesses. They suspected that hereditary factors—what they called genes—were part of the answer. But they couldn't say what genes were.

It was possible, in fact, that a gene was nothing but an abstraction. Heredity's patterns might emerge from a combination of certain subtle features inside cells. But as scientists looked more closely at cells, they came to suspect that genes had something to do with mysterious, thread-shaped objects known as chromosomes. They could see the twenty-three pairs of chromosomes in each of our cells. They could watch them double into two sets when the cell divided. They could observe sex cells end up with just one copy of each chromosome, to be joined to the other copy at fertilization. But no one knew what controlled these movements, or how inherited chromosomes might influence our traits.

Trying to describe this dance in 1923, the biologist Edmund Wilson confessed that it was so intricate that he found it hard to accept as real. "We find ourselves fairly gasping for breath," Wilson admitted. "Such results are indeed staggering—to a certain type of mind even harder to assimilate than those which physicists are now asking us to accept concerning the structure of atoms."

Some of the most important clues about chromosomes and heredity came from a room full of flies at Columbia University. There a biologist named Thomas Hunt Morgan led a team of scientists who examined the chromosomes of fruit flies under microscopes. Chromosomes have bands along their length, like cellular snakes. By tracing the molecules from one generation to the next, they could track pairs of chromosomes as they traded sections with each other.

Morgan's team proved that inheriting a short piece of a chromosome could determine a trait in a fly. It could set the color of a fly's eye to red or white. It could make a fly withstand the cold or freeze to death. These drastic results made Morgan suspect that genes lurked in these chromosomal segments.

He couldn't say much more. For one thing, chromosomes were a hideous biochemical mess, a mash-up of proteins and a particularly strange substance known as nucleic acid. But one of Morgan's students, Hermann Muller, gained a crucial clue about genes by blasting flies with X-rays. Every now and then he produced a mutation in a fly—a fly whose ancestors all had red eyes suddenly developed brown ones, for example. If Muller then bred a mutant fly, it could pass down the new trait. In other words, he had changed a gene.

Muller suspected that mutations happened on a regular basis. Nature didn't need an X-ray machine to change genes. High temperatures or certain kinds of chemicals might randomly alter a gene every now and then. And from such blind changes, all of life's variations arose. In 1926, Muller declared that the gene was "the basis of life."

In 1932, Muller came to Berlin to work with geneticists there, trying out different kinds of radiation on flies to see what sort of mutations they might create. Delbrück was awestruck when he met Muller and decided he would bring his knowledge of quantum physics to bear on the phenomenon. After Muller left Berlin for a job in the Soviet Union, Delbrück began collaborating with the geneticists, carrying out what he called his "black market research."

Delbrück recognized that genes, whatever their exact nature, were deeply paradoxical. They were stable enough to be passed down for thousands of generations, only to abruptly mutate and then become stable once more. Delbrück saw a solution to the paradox in physics.

If an atom absorbs a photon of light, one of its electrons may jump to a higher energy level, where it will remain. X-rays might have the same effect on a gene. The fact that X-rays, with their exquisitely narrow beams, could cause a mutation also meant that genes must be exquisitely small.

"These are primarily speculations," Delbrück and his colleagues warned in a 1935 paper, "which rest on still shaky ground."

If they were worried that their paper would trigger a wave of misconceptions, they would have been relieved. It appeared in a journal that Delbrück later said nobody read. Their ideas, he said, "got a funeral first class."

Soon after the publication of his gene paper, Delbrück escaped Nazi Germany. He left behind not just his country but his science. Giving up physics to become a full-blown biologist, Delbrück made his way to the laboratory of Thomas Hunt Morgan, who was now working at Caltech. But once he got there, Delbrück felt that he had made a terrible mistake. Morgan "didn't know what to do with this theoretical physicist," Delbrück later recalled. And when he tried to run experiments on Morgan's flies, he found the work tedious and the results inscrutable.

In a stroke of good fortune, Delbrück ran into a biochemist named Emory Ellis one day. He was intrigued to discover that Ellis studied phages instead of animals. The experiments Ellis ran were simple yet powerful. He added bacteria-killing viruses to petri dishes, where he could see ghostly holes form where the phages killed millions of their hosts. He needed only transfer a bit of the agar from a hole into an uninfected dish to unleash a new outbreak. The phages appeared to have genes of their own, but they reproduced simply by making copies of themselves. There were no messy blendings of chromosomes to struggle through.

Delbrück fondly called viruses the "atoms in biology." He began running experiments of his own, which soon blossomed into work that would later earn him the Nobel Prize. Viruses mutated like flies, it turned out. Some mutations took away their power to infect a strain of bacteria; others let them attack a new one. By counting the ghostly holes in his petri dishes, Delbrück could make precise measurements of how often mutations arose. He was happy in his new incarnation, even if few people at the time recognized the new kind of science he was building.

In 1945, a few years into his new career, a friend handed Delbrück a slender new book that was all the rage. The title was *What Is Life?*, and it left Delbrück gobsmacked. The author was Erwin Schrödinger, a physicist whom Delbrück knew back in his old quantum physics days in Germany. And to answer the question that formed the book's title, Schrödinger had resurrected the paper of Delbrück's that had gotten a funeral first class.

Erwin Schrödinger was born in Vienna in 1887 and went on to become a physics professor in Zurich, where he developed an equation that would bear his name. The Schrödinger equation predicts how a system—be it a photon or an atom or a group of molecules—changes through time and space in a wavelike fashion. But Schrödinger's name also became affixed to the most famous thought experiment involving a cat.

Schrödinger recognized the profound weirdness that his work and that of other quantum physicists implied. He offered a way to picture that weirdness: Think of a cat in a box. The box is rigged up with a device that can flood it with poison and kill the cat. Now imagine that the device can activate in response to a radioactive atom spontaneously decaying.

According to the leading interpretations of quantum physics in the 1930s, the atom could exist in a decayed state and an undecayed one at the same time. Only an observation would force its wavelike nature to collapse into one state or the other. If quantum physics was correct, Schrödinger argued, the cat had to be at once dead and alive. Only when an observer looked in the box did the cat receive just one fate.

To Schrödinger, life and death were more than just fodder for thought experiments. His father, a botanist, had introduced him as a boy to the complexities of plants. As a university student, he devoured biology books. Later, when Muller created mutations with X-rays, Schrödinger became intrigued by the nature of genes. He developed a layman's curiosity about "the fundamental difference between living and dead matter," as he once put it. When a friend passed Delbrück's 1935 paper about genes to Schrödinger, it became the nucleus around which his own thoughts grew. At the time, Delbrück and Schrödinger were professional colleagues who traveled in Europe's rarefied circles of quantum physicists. But Schrödinger never once spoke or wrote to Delbrück about the inspiration he provided.

Like Delbrück, Schrödinger sought refuge from the Nazis. Instead of California, he ended up in Ireland, where the government built him a research center to run. One of the requirements of the job was a series of public lectures at Trinity College. Schrödinger decided not to talk about his equations, since he would not be speaking to an audience of quantum physicists. Instead, he would deliver a course of lectures on his private thoughts about the nature of life.

A vast crowd descended on the lecture hall in February 1943. The organizers had to turn thousands away. When Schrödinger rose to speak, he warned the packed hall that he spoke not as an expert but as "a naive physicist." And he had a naive question to ask—the same one that Georg Stahl had asked nearly 250 years earlier: What is life?

Much of the biology that Schrödinger described to his Dublin audience was not new. And much of what was new in his lectures would eventually prove to be wrong. And yet he managed to frame much of modern science's approach to what it means to be alive. His ideas guided a generation of scientists who put biology on a molecular footing. And, just as importantly, he made clear to physicists just how badly their theories failed when they crossed into life's territory. Eighty years later, they're still struggling to meet his challenge.

"What an organism feeds upon is negative entropy," Schrödinger declared. Entropy is essentially a measurement of disorder. The jostling of atoms and molecules naturally increases entropy over time. For life to maintain its order, it needs to draw in energy in a way that counteracts the rise of entropy. And it extends its order into the future by passing on its genes to its descendants.

To explain heredity, Schrödinger relied on Delbrück's work a decade earlier on chromosomes. Schrödinger envisioned them as stable crystals that contained genes, and that could be replicated from one generation to the next.

Schrödinger had only the vaguest idea how this arrangement would work. The crystals would have to be what he called "aperiodic." Ordinary crystals repeat themselves in periodic patterns: ice is a latticework of water molecules, table salt a series of cages made of sodium and chloride. No matter where you move around inside these crystals, the pattern is identical. But chromosomes, Schrödinger speculated, have arrangements of atoms with variations that do not merely repeat—like a string of letters chosen from an alphabet. Those variations could serve as a "code-script," as Schrödinger called it, that could produce an entire organism.

"The difference in structure," Schrödinger speculated, "is of the same kind as that between an ordinary wallpaper in which the same pattern is repeated again and again in regular periodicity and a masterpiece of embroidery, say a Raphael tapestry, which shows no dull repetition, but an elaborate, coherent, meaningful design traced by the great master."

Despite his musings on entropy and code-scripts, Schrödinger's lectures proved immensely popular—so popular, in fact, that he had to deliver them in full a second time. As reports of his sensational ideas spread, a publisher invited him to write them up as a short book. *What Is Life?* became a hit the following year. The book didn't just fascinate the public; it also steered the course of science. Within nine years of its publication, two readers of *What Is Life?* would discover that Schrödinger's aperiodic crystal was not just an idea but a real molecule: the DNA nestled in each of our cells.

One of those readers was an English physicist named Francis Crick. Born in 1916 to middle-class parents in suburban England, Crick lost his faith by his early teens, turning instead to science to understand the world. The mysteries that science had yet to explain, he later wrote, "serve as easy refuge for religious superstition." Crick chose to study physics at University College London but did not impress his teachers. In graduate school he was assigned the task of measuring the viscosity of water, which he called "the dullest problem imaginable."

Crick spent World War II working at Britain's Admiralty Research Laboratory, designing underwater mines to sink Nazi ships. When peace came, Crick didn't want to go back to the viscosity of water, nor did he want to build more war machines. He craved something profound. One day he happened to read about the recent discovery of antibiotics, and the idea that these molecules could save people's lives left him electrified. When Crick told his friends about them, he was struck by the enthusiasm in his voice. He wondered if, at age thirty, he could make a radical shift to become a biologist.

It was around that time that he read *What Is Life?* Schrödinger gave Crick an infusion of confidence that the shift to biology might not be so radical after all. Life was just a part of the world that physics had yet to explain very well. Crick decided he wouldn't limit his focus to a single antibiotic or some other organic molecule. He was attracted instead to what he called "the borderline between the living and the nonliving."

Crick's hostility to religious superstition helped push him in that direction, too. His boyish contempt for the church expanded in adulthood. He despised intellectuals who claimed that life would always defy a reduction to simple mechanisms. To Crick, they were just leftover vitalists. Even after World War II, the French philosopher Henri Bergson remained trendy, while the theologian and paleontologist Pierre Teilhard de Chardin rose to fame for claiming that molecules were infused with purpose, giving rise first to life and ultimately to consciousness. In England, the writer C. S. Lewis cast scorn on modern science's bleak vision of the world, hoping it would be replaced by an investigation of nature that did not destroy life's glories. "When it explained it would not explain away," Lewis said in 1943. "When it spoke of the parts it would remember the whole."

For Crick, the only way to understand the whole was to start with the parts. He landed a spot at the Cavendish Laboratory, the same institution where John Butler Burke had been fooled by radiobes some four decades earlier. In the early 1900s, Burke's obsession with life made him stand out as an oddity at the Cavendish. All of his colleagues were content to study electrons, radioactivity, and other lifeless things. By the 1940s, however, the Cavendish physicists were using their expertise to make sense of biological molecules.

To figure out the structure of life's compounds, they coaxed the molecules to assemble into crystals. The scientists then blasted them with X-rays, which glanced off the crystals and crashed into photographic plates. The ghostly spots and curves that formed in the photographs hinted at the repeating structure of the crystals.

The Cavendish researchers could then use mathematical equations to work their way back from the marks in the photographs to the shapes of the molecules. They started simply, with vitamins and other small molecules, and moved on to the daunting challenge of proteins, which were huge snarls of amino acid chains.

Working out the structures of proteins would help scientists understand how they behaved and what jobs they served. Biochemists were already somewhat familiar with enzymes, which were proteins that sped up chemical reactions. Other proteins seemed to serve as signals, and others locked together like bricks for the body's edifice. In the 1940s many biochemists suspected that genes were made of proteins nestled in the chromosomes.

After Crick arrived, he quickly impressed the scientists at the Cavendish with his preternatural ability to picture the twists and folds of proteins in his mind and see what their X-ray portrait would look like. But not long after he started on this work, Crick became distracted. A series of experiments in the late 1940s and early 1950s showed that proteins were not, after all, the carriers of hereditary information. DNA, a nucleic acid also found in the tangle of chromosomes, proved essential.

At the time no one knew much about DNA's structure. Crick mused about what sort of shapes DNA would need to act as Schrödinger's aperiodic crystal. His bosses at the Cavendish discouraged him from his daydreaming, but in 1951 he met a young American visiting Cambridge who also loved *What Is Life?* James Watson was happy to talk about DNA with Crick for hours on end.

Their conversations could only take them so far, though. If DNA was the code-script of life, they wanted to know how it stored genes. Crick and Watson knew that a team of scientists in London was trying to create the first good pictures of DNA crystals. Led by Rosalind Franklin, they worked carefully and methodically to prepare the molecules, bombed them with X-rays from different angles, and inspected the images the beams produced.

Franklin had no tolerance for Crick and Watson's impatience, once even ejecting Watson from her lab so she could get back to her work. When they tried building a model based on some preliminary images, she traveled to Cambridge and explained to them that it was all wrong. Later, unbeknownst to her, Crick and Watson got to see some of her unpublished work. Those clues were finally enough for them to come up with a new structure that they believed could fit what scientists knew about DNA's chemistry and even explain how DNA could serve as the stuff of genes.

Compared to proteins, with their maddening switchback curves and intertwining clumps, DNA was graceful in its simplicity. Crick and Watson recognized that DNA was a pair of twisting backbones with rung-like connections linking them together. Each rung was composed of a pair of compounds, called bases. At each rung in a piece of DNA, a base can take one of four forms. Each gene, stretching for thousands of base pairs, is a unique sequence of these bases.

"Now we believe that the D.N.A. is a code," Crick wrote his twelve-year-old son Michael in 1953. "That is, the order of the bases (the letters) makes one gene different from another gene (just as one page of print is different from another)."

Crick and Watson's model also showed how living things could maintain the order of their genes. A cell could make a copy of its DNA by pulling its two backbones apart, each with a set of bases still dangling from it. Each kind of base can only bond with one other kind of base, making it easy for two accurate copies to take shape.

On the day that Crick and Watson realized they had worked out the structure of DNA, they marched to a nearby pub to celebrate their breakthrough. Crick shouted that they had "found the secret of life." It was a cry of victory in his war against the vitalists. Along with Franklin and their other colleagues, they wrote up their results, and the batch of papers appeared in *Nature* on April 25, 1953, laying out the double-helix model and the evidence for it. When the *New York Times* interviewed Crick, he said that, for now, the idea "simply smells right."

In August Crick sent reprints to Schrödinger in Dublin, with a short note: "You will see that it looks as though your term 'aperiodic crystal' is going to be a very apt one."

DNA did not immediately become the icon of life that it is today. When Crick and Watson shared the Nobel Prize in 1962, it gained some fame. (Franklin could not be considered for the award because she died of cancer in 1958.) But the molecule only penetrated pop culture when Watson published a best-selling account of the discovery in 1968, *The Double Helix*. In its pages he reprinted a photograph taken of himself and Crick shortly after their paper came out in *Nature*.

The two scientists posed in their Cavendish lab next to a man-sized model of the double helix they had built from rods and plates and screws. Watson looked on as Crick pointed to a twist in the backbones with a slide rule. That photograph came to represent the turning point in our modern conception of life. One historian ranked it as one of the most important photographs of twentieth-century science, along with Einstein's portrait and the image of a mushroom cloud.

But icons inevitably distort history. Rosalind Franklin is missing from the photograph, for one thing. And the picture also does a disservice to Crick's memory. It traps him in the frame with the double helix, as if that were his sole accomplishment. But Crick went on to do work that proved to be just as profound: he joined forces with an international network of scientists to figure out the rules by which cells translate the information in genes into the structure of proteins. They named the rules the genetic code. No one photographed Crick pointing a slide rule at the code. Yet its discovery was arguably just as important as that of the double helix.

Genes and proteins, Crick recognized, are spelled out in different alphabets. DNA is made out of four different bases. Proteins, on the other hand, are assembled from about twenty different amino acids. Once our cells make an RNA copy of a gene, they feed it into a protein-making factory called a ribosome. Crick and his colleagues figured out that the ribosome reads three bases in a row in order to determine which amino acid to add to a protein. If a mutation changes one of those bases, it may result in cells building a protein with a different amino acid in that position.

For Crick, the genetic code didn't just smell right; it represented the triumph of his scientific approach to life. "It is, in a sense, the key to molecular biology," he declared. "It will be difficult, after this, for doubters not to accept the fundamental assumptions of molecular biology which we have been trying to prove for so many years."

But Crick could not stay gracious in victory. Much to his dismay, the discovery of the genetic code failed to make the vitalists see the error of their ways. All around him, Crick saw vitalists on the rise. One day a Cambridge clergyman informed him that DNA might be evidence for extrasensory perception. Crick read with horror of a Princeton physicist named Walter Elsasser, famous for working out how Earth generates a magnetic field, who decided to try his hand at biology. In 1958, Elsasser claimed to discover "biotonic phenomena" that "cannot be explained in terms of mechanistic functioning." Another scientist wrote to the journal *Nature* to claim that what distinguished the living from the nonliving was a "biological urge" that could never be explained with atoms and molecules.

Crick got so exasperated that he began giving lectures around Cambridge warning of the threat vitalism posed to civilization. Soon afterward the University of Washington asked him to give a series of lectures in Seattle about the impact of science and philosophy on "man's perception of a rational universe." Crick used the opportunity to deliver a high-profile attack on his enemies. He entitled his talks, "Is Vitalism Dead?"

In Seattle, Crick regaled his audience with the dizzying advances he had been a part of: making sense of heredity, the genetic code, the workings of cells. In spite of all that evidence, he ruefully observed, vitalism still lingered. Crick blamed its endurance on our weakness for superstitions. The only solution Crick could think of was to take over the schools. To counteract the illusions fostered by the arts, all students should be required to take a heavy load of science classes. The old literary culture was "clearly dying," Crick declared, and would be replaced by a new culture based on "science in general, and natural selection in particular."

Crick closed his Seattle lectures with a harsh warning. "And so to those of you who may be vitalists I would make this prophecy: what everyone believed yesterday, and you believe today, only cranks will believe tomorrow."

As adept as Crick might be as a scientist, he turned out to be a clumsy polemicist. When his lectures were published in 1966 as the book *Of Molecules and Men*, one reviewer called it "a frightening mixture of naïveté and bigotry."

It was so bad that many of his toughest critics were his fellow scientists. Vitalism had spiraled into scientific oblivion by the 1930s. The embryologist Conrad Waddington wondered if Crick was just "flogging a dead horse." Sir John Eccles, a leading neuroscientist, praised the parts of his lectures where Crick described the new science of molecular biology. But he dismissed Crick's science-centered vision of society as a "dogmatic religious assertion." Eccles also took Crick to task for crudely dismissing anything that lay beyond atoms and molecules as vitalism.

"In biology," Eccles argued, "there are new emergent properties not predictable from chemistry, just as chemistry is not predictable from physics."

Crick would never vanquish his vitalists. Part of his problem was that he used the term as a broad-brush insult against anyone who didn't see things his way: a motley crew of novelists, fans of extrasensory perception, and even some full-time scientists. But another part of the problem lay in his own work on DNA and the genetic code. As profoundly important as it was, it still left many big questions unanswered about what set the living apart from the nonliving. In 2000, four years before Crick's death, a trio of leading biologists published a review they called "Molecular Vitalism." They argued that a simple, machine-based view of life with DNA serving as instructions was not powerful enough to explain some of the most important features of the living world: how cells remain stable in a world of flux, for example, or how embryos reliably develop into complex anatomies. Looking over the gene-centered state of biology at the turn of the millennium, they doubted that it would "convince a nineteenth century vitalist that the nature of life was now understood."

As Crick got older, he indulged in reflections on life that a young scientist would rarely dare. He mused about aliens. Perhaps they were like us. Perhaps they even seeded Earth with life in the first place. Or perhaps aliens were chemically different from life as we know it. Perhaps they lived on a gas planet, or inside a sun. No matter how strange alien life might turn out to be, though, Crick suspected that it would probably turn out to be a lot like life on Earth. There existed something Crick called "the general nature of life."

Crick sketched out that general nature in 1981. "The system must be able to replicate directly both its own instructions and indirectly any machinery needed to execute them," he wrote in his book *Life Itself.* "The replication of the genetic material must be fairly exact, but mutations—mistakes which can be faithfully copied—must occur at a rather low rate. A gene and its 'product' must be kept relatively close together. The system will be an open one and must have a supply of raw material, and, in some way or another, a supply of free energy."

Only fifty years had passed since Max Delbrück had contemplated the riddles of life in his furtive Berlin get-togethers. Now his intellectual grandchildren thought about life largely in terms of genes—the ability that genes had to encode the molecules necessary to copy themselves and the power they had to drive evolution. Crick's work had a profound impact on how other scientists defined life. In 1992, for example, his influence could be felt at a meeting that NASA organized to come up with ideas for how to study the possibility of life on other worlds.

One of the scientists at the meeting, Gerald Joyce, later described the meeting to me. "We're talking about the search for life and the origin of life," he recalled, "and someone said, 'Do you think we should actually define what it is we're talking about?'"

The scientists started throwing out ideas, shooting some down and merging others together. The conversation started at the official meeting and lasted through dinner. Like Crick, the NASA group saw metabolism as essential—but mostly because it provided the material and energy to make new copies of genes. Life couldn't copy those genes perfectly, however. Only if it made mistakes could evolution emerge and allow life to adapt and take on new forms, which could then be passed down through the generations. "History starts to be written in molecules," Joyce told me later. "That's why biology is different than chemistry."

By the end of dinner the scientists had distilled their ideas down to eleven words:

"Life is a self-sustained chemical system capable of undergoing Darwinian evolution."

Crisp and concise, short enough to memorize, their wording took hold. People started referring to it simply as "the NASA definition of life," as if the space agency had given it an official stamp of approval. At scientific conferences, speakers flashed those eleven words in their slide decks. They made their way into textbooks. Reading the definition, students could be forgiven for assuming that the whole matter was settled.

Far from it. Like its predecessors, the NASA definition of life did not come with a list of the things that qualified as life and the ones that didn't. And when scientists turned to the real things with which we share the world, they couldn't agree on what belonged.

PART FOUR RETURN TO THE BORDERLAND

HALF LIFE

"Mr. Burke is not prepared to affirm with positiveness that these organisms are quite alive. They may be *half alive.*"

In the spring of 2020, coyotes strolled down the daytime streets of San Francisco. Pods of pink dolphins cavorted in the waters around Hong Kong. A herd of mountain goats took over a town in Wales, and jackals wandered a city park in Tel Aviv. In Venice, cormorants plunged into the suddenly clear canals to chase after fish, while Canada geese escorted their goslings down the middle of Las Vegas Boulevard, passing shuttered-up shops selling Montblanc pens and Fendi handbags.

A strange expansion of life was taking place, thanks to the retreat of our species. Billions of people went into lockdown for months, a movement that scientists dubbed the anthropause. For the lucky ones, the greatest challenge of this retreat was boredom. For the unlucky, unemployment, hunger, and other disasters awaited. For the most unlucky of all, there was sickness. Their bodies flared with fever and shook with raw coughs. Some of the sick shivered at night so violently they chipped their teeth. Four out of five sickened people rode out the disease at home. One out of five ended up in the hospital. The lungs of some became wastelands of pus and inflammation. Hundreds of thousands died. In New York City, backhoes dug trenches on Hart Island to bury the overflow of coffins.

The new pneumonia first came to light in late 2019 in the Chinese city of Wuhan. Within a few weeks Chinese researchers had isolated the microscopic thread that tied all the cases together: a virus, which virologists named SARS-CoV-2. They analyzed its genes and reconstructed parts of its history recorded in mutations. SARS-CoV-2 arose from bats, as have a number of other dangerous viruses in recent decades. And like those viruses, SARS-CoV-2 evolved adaptations that let it thrive inside humans instead.

A cough or even a song could loft a spray of virus-laden droplets into the air, ready to be inhaled by someone riding the same bus, sharing the same breakfast table, praying in the same church. Once inside a new nose, the virus could infect a new host. The virus was studded with proteins that could latch onto a protein on the surface of certain cells in the airway. Its membrane fused with the cells, and it dumped its genes inside. They were spelled out according to the same genetic code our cells use to build proteins. As a result, the cells translated the virus's genes into proteins as they would their own. Now the cells filled themselves with viral proteins that shut down their ordinary work and forced them to make new viruses. They made new copies of the virus's genes, which were then cradled inside new protein-studded membranes. The new viruses gathered into bubbles that migrated to the border of the infected cells and spilled open, delivering millions of new viruses into the airway.

In most people the immune system got wind of the invasion while the viruses were still establishing themselves in their noses. They mounted a defense, learning how to deliver a precise attack with antibodies that could stop the viruses from infecting new cells. But the viruses had their own cunning evasions encoded in their genes. They could silence the alarm systems inside the cells they invaded. In some people, they proliferated beyond control, working their way down into the lungs. The immune system lost its surgical precision, resorting to brute-force attacks, spewing toxic compounds in all directions. The virus's victims slowly drowned in an ocean of their own making.

If SARS-CoV-2 always laid its victims so low, it might have been easier to battle. When people became ill, they could be put into quarantined hospital rooms. But SARS-CoV-2 lurked quietly in its hosts for days before creating its first symptoms. People went about their lives unaware of the multiplying viruses inside them or the clouds of infection they exhaled. They lingered over lunch at restaurants, they worked at call centers, they leaned on the railings of cruise ships plowing the Pacific. After infecting people around them, some of the virus's hosts finally developed symptoms. Others never did.

The unwittingly infected exported Covid-19 out of Wuhan. Some traveled across China to celebrate the Lunar New Year with their families. Planes delivered infected passengers to Europe, and from there to other continents. The virus mutated as it multiplied, and new lineages emerged, marked by different gene signatures. Scientists reconstructed their journeys from their mutations as they moved between countries and among cities. Some nations managed the pandemic well, while others—due to the limits of poverty or the arrogance of wealth—suffered its full ravages.

It is hard to think of something that has laid such a heavy blow on humanity in so little time. It is hard to think of something that has done a better job at reproducing, using our species to make quadrillions of copies of itself in a matter of months.

And yet, for all that, there are many scientists who would say that SARS-CoV-2 is not alive. It does not deserve entry into the exclusive club called Life.

For thousands of years, people knew of viruses only through the death and destruction they caused. Doctors gave their diseases names, like smallpox, rabies, and influenza. When Antonie van Leeuwenhoek peered at drops of water with his microscope in the 1600s, he discovered bacteria and other minuscule wonders, but he could not see the even tinier viruses. Two centuries later, when scientists finally discovered viruses, they did so without actually seeing them.

In the late 1800s a handful of scientists in Europe studied a disease of tobacco plants called tobacco mosaic disease. It stunted the plants and covered their leaves with spots. Mashing up a sick leaf in water, the scientists injected the fluid into healthy plants and watched them get sick, too. But when they searched for the pathogen in the fluid, they couldn't find bacteria or fungi. It had to be something fundamentally different.

A Dutch scientist named Martinus Beijerinck poured mashed-up leaves from a sick tobacco plant through a porcelain filter. The pores were too small for any bacteria to sneak through. He was left with a clear liquid. But that was enough, when injected into a new plant, to pass on the disease. Beijerinck concluded that some invisible agent multiplied in tobacco plants. In 1898 he dubbed it a virus, using an ancient word for toxin.

Virologists went on to find the viruses that caused rabies, influenza, polio, and many other dread diseases. Some viruses infected certain species of animals, while others infected only plants. Biologists discovered phages, the viruses that only infect bacteria. And it was a phage that became the first virus ever seen by humans.

In the 1930s engineers built electron microscopes powerful enough to bring the viral world into focus. The device revealed phages sitting atop a bacterial host. They looked like crystals set on leglike wires. Other viruses turned out to resemble serpents; others, soccer balls. SARS-CoV-2 belongs to the coronaviruses, named for the halo of proteins that adorn their surface. They reminded virologists of a solar eclipse, when the sun's corona of streaming gas becomes visible.

Biochemists broke down viruses into their molecular ingredients. They started with Beijerinck's tobacco mosaic virus, finding that it contained proteins built from the same set of amino acids as our own. But among those proteins biochemists could not find any of the enzymes that our own cells use for metabolism. Viruses do not eat or grow. Old viruses do not beget new ones, at least not directly. A virus is just a reorganized package of its host's own atoms.

For biologists searching for a definition of life, viruses became a headache. They couldn't dismiss viruses altogether, because they clearly had some of life's hallmarks. And yet they lacked others. It would have been convenient if viruses had turned out to be mirages like *Bathybius* or radiobes. But the more that scientists studied viruses, the more real they proved to be—and the more perplexing their nature.

"When one is asked whether a filter-passing virus is living or dead," the British virologist Norman Pirie wrote in 1937, "the only sensible answer is: 'I don't know; we know a number of things it will do and a number of things it won't and if some commission will define the word "living" I will try to see how the virus fits into the definition.'"

Pirie and his fellow virologists went on to discover crucial features of viruses. Inside their protein shells and oily membranes, they contain bundles of genes, along with some proteins to hold them together. But they contain none of their own ATP to fuel reactions. On the outside, viruses have a furry coat of sugar-frosted proteins. The proteins typically fit precisely onto proteins on the surface of cells. This latching is the first step of a virus infection, and it has to be a precise fit, like a key in a lock. That's one reason why viruses are so selective about the species they infect and why they can invade some types of cells but not others.

Once a virus enters a cell, its shell or membrane breaks apart and it delivers its payload of genes. If the copying of genes is the crux of life, then viruses should certainly qualify as living. Some viruses have genes encoded in DNA, using the same four-letter alphabet that spells out our own heredity. An infected cell will read that viral DNA and make RNA molecules, which it can then turn into proteins for the virus.

But Pirie and other virologists discovered that many viruses have streamlined this transformation. In the 1930s, Pirie found hints that the genes of tobacco mosaic viruses were made not of DNA but of RNA. Later research revealed that many other viruses use RNA for their genes, including SARS-CoV-2. When RNA viruses invade a cell, their genes get translated straight to proteins. It is an exquisitely efficient way for viruses to make us sick for their own benefit. And yet only viruses have discovered this particular kind of biochemistry.

Whether viruses use DNA or RNA to encode their genes, they can get by with astonishingly few. We carry 20,000 protein-coding genes. SARS-CoV-2 was able to hurl the global economy into an abyss with only twenty-nine. Each time SARS-CoV-2 invades a cell in someone's airway, the millions of new viruses that come out bear those twenty-nine genes, usually in identical form. But some bear mistakes.

Viruses mutate like more familiar forms of life. In fact, they mutate at a far higher rate than humans, plants, or even bacteria. Our cells contain a molecular staff of proofreaders checking new sequences of DNA for errors and sending most mistakes back for fixing. Most viruses can't check for errors. SARS-CoV-2 and other coronaviruses are peculiar because they carry a gene for a primitive proofreading protein. Even though they don't mutate as fast as most other viruses, they still build up mutations thousands of times faster than we do.

Sometimes those new mutations give a virus a competitive edge over other viruses. They may speed up the time it takes to replicate. They may enable a mutant virus to become invisible on the immune system's radar. These viruses will be favored by natural selection.

The modern study of viruses has revealed, in other words, that viruses share yet another hallmark of life: evolution. They can evolve resistance to antiviral drugs. They can evolve to adapt to a new host species. Evolution figured prominently in the NASA definition of life, and yet Gerald Joyce, one of its architects, didn't think the evolution of viruses was enough to make up for the fact that they are not a self-sustained chemical system. Viruses get their sustenance inside the chemical system of a cell, and only inside a cell can they evolve.

"According to the working definition, a virus doesn't make the cut," Joyce decreed in an interview with *Astrobiology Magazine.*

Viruses have had their defenders, though. Starting in 2011, the French scientist Patrick Forterre made a series of arguments in favor of viruses as being alive. At least, they're alive some of the time, he said. To Forterre, the cell is the fundamental feature of life. And when a virus invades it, the cell effectively becomes an extension of the virus's genes. Forterre likes to call it a virocell.

"Whereas the dream of a normal cell is to produce two cells, the dream of a virocell is to produce a hundred or more new virocells," he wrote in 2016.

Forterre did not win over many of his fellow virologists. Purificación López-García and David Moreira called his argument "alien to logic." Others dismissed the virocell as mere poetic license. Viruses can no more live than they can dream. And when the International Committee on Taxonomy of Viruses established a modern system of classification, they flatly declared that "viruses are not living organisms."

"They lead only a kind of borrowed life," one committee member explained.

It's strange that people can push viruses out of the house of life and leave them hanging around the doorstep. It's awfully crowded out there. There are more viruses in a liter of seawater than there are human beings on the entire planet. The same is true for a spoonful of dirt. If we could count up all the viruses on Earth, they would outnumber every form of cell-based life combined, perhaps by a factor of ten.

The diversity of viruses is also colossal. Some virologists have estimated that there may be trillions of species of viruses on the planet. When virologists find new viruses, they're often from a major lineage no one knew about before. Ornithologists get justifiably excited when they discover a new species of bird. Imagine what it would be like to discover birds for the first time. That's what it's like to be a virologist.

Can we exile all this biological diversity from life? To exile viruses also means we have to discount how intimately woven they are into life's ecological web. They rival predators in their slaughter, whether they are killing off a coral reef or wiping out *Pseudomonas* in a lung. Viruses also have peaceful relationships with many of their hosts. Our healthy bodies are home to trillions of viruses collectively known as a virome. Most of them infect the trillions of bacteria, fungi, and other single-celled members of our microbiome. Some studies suggest that the human virome keeps our microbiome in balance, contributing to our own well-being.

The earth has a virome of its own, one that acts as a geochemical force. Each time you blink an eye, 10 billion trillion phages in the ocean infect marine bacteria. Many of them kill their microbial hosts, dumping about three gigatons of organic carbon into the water every year, stimulating the growth of new life. Some phages are more merciful: they slip inside their hosts and let them go on with their lives for a while. Some even bring genes with them that help their hosts thrive. There are phages that float from host to host in the ocean with genes for photosynthesis. The microbes they infect do better at harnessing sunlight. The oxygen we breathe is brought to us in part by these viruses.

These phages came by their light-harvesting genes by theft. When their ancestors infected other photosynthetic microbes, they accidentally folded their host's genes into their own as they replicated. But viruses can also donate new genes to the genomes of their hosts. Bacteria can gain resistance to antibiotics through a viral infection, for example. Our own genome contains tens of thousands of viral fragments, adding up to 8 percent of our DNA. Some of those fragments have evolved into genes and switches for turning genes on and off. If viruses are lifeless, then lifelessness is stitched into our being.

Viruses are not the only things that straddle life's edge. Think of the red blood cells that course through your veins. You'd be dead without them, starved of the oxygen they ferry from the lungs throughout the body. Red blood cells (also known as erythrocytes) have membranes, just as bacteria and slime molds do.

Inside, they are full of sophisticated enzymes and other proteins. Red blood cells even get old and die. "The life span of erythrocytes amounts to some 100–120 days," a team of scientists reported in a 2008 review. If something has a life span, surely it has a life.

And yet, by many definitions, red blood cells aren't alive, either. Unlike other cells in our bodies, they have a peculiar path of development. They arise from precursor cells in our bone marrow and then get released into the bloodstream. They take with them the hemoglobin and other proteins they will need to carry oxygen. But they don't take any DNA. As a result, a mature red blood cell lacks the genetic cookbook to make its own proteins and divide into new cells.

Red blood cells are different from other cells in another important respect: they cannot make their own fuel, because they lack the factories for making it. Other cells contain dozens of free-floating bags of enzymes called mitochondria. And it turns out that mitochondria, too, are a form of half life. Each mitochondrion carries thirty-seven of its own genes, along with ribosomes it uses to make proteins from them. And from time to time a mitochondrion will multiply the way bacteria do, pinching itself off down the middle and becoming two new mitochondria, each with its own circle of DNA.

The solution to the puzzle of mitochondria lies deep in our history. Two billion years ago the ancestors of our mitochondria were free-living bacteria. They were engulfed by a larger cell, and the two species formed a partnership. In exchange for ATP the mitochondria got shelter. No longer required to survive on their own, mitochondria lost most of their genes—but not all of them. And they have not lost the ability to divide as their bacterial ancestors did.

Run down a typical list of requirements for life, and mitochondria have most of them—more than red blood cells, in fact. Yet they cannot exist outside their host cells. They cannot find their own food. They cannot build their genes or proteins on their own. They were certainly once living things, but now it's hard to say what they've become. Calling them dead certainly doesn't seem right, since our own lives depend on them.

Still, mitochondria and red blood cells are so small that perhaps we can ignore them. Out of sight, out of mind. But some of life's paradoxes are not invisible to the naked eye. In 1948, Albert Szent-Györgyi slyly observed that if life were characterized by self-reproduction, then a single rabbit was not alive. After all, a single rabbit cannot make more rabbits. Many scientists have ignored Szent-Györgyi's warning, judging from the fact that they've made self-reproduction a requirement for life. We can be charitable and assume the people who made these definitions think Szent-Györgyi was merely playing word games. It doesn't matter if one rabbit can't reproduce because it belongs to a species that can.

But nature, it turns out, causes more trouble than even Szent-Györgyi could.

In the 1920s a husband-and-wife team of naturalists named Carl and Laura Hubbs traveled around Mexico and Texas catching fish. They got to know the animals in intimate detail, down to their stripes, spots, and rays. This loving, encyclopedic attention revealed to them that many species of freshwater fish evolved through interbreeding. Two species interbred, and their hybrid offspring could now only mate among themselves. But one of these hybrid species, a relative of guppies called *Poecilia formosa*, proved remarkably different from the others.

"Not a single male has been found, among about two thousand

specimens examined from Tamaulipas and Texas," the Hubbses reported. They nicknamed the fish Amazon mollies—not for the river but for the female warriors of ancient tales.

Amazon mollies evolved about 280,000 years ago from the interbreeding of two species of fish: the Atlantic molly and the sailfin molly. Once the new species evolved, it never left its parents. Today Amazon mollies are always found alongside either Atlantic mollies or sailfin mollies. It is as if the survival of the species depended on their company.

To make sense of these patterns, the Hubbses brought all three species back to their laboratory at the University of Michigan. They dropped the fish into tanks and let nature take its course. The female Amazon mollies mated with both Atlantic and sailfin males. They laid eggs, out of which Amazon mollies always hatched. And, true to their name, all those Amazon offspring were daughters.

"Although the broods have been large and many," the Hubbses observed, "not a single male has appeared among them."

In the mid-1700s, Abraham Trembley observed that female aphids could reproduce without males, producing a line of daughters and granddaughters. In later generations others found more invertebrates that could perform this kind of virgin birth, known as parthenogenesis. Their eggs spontaneously developed into embryos without any need for a male's sperm. And when the Hubbses investigated Amazon mollies some two centuries later, they discovered that vertebrates can be parthenogenetic, too.

Unlike aphids, however, Amazon mollies need to mate with males. As later experiments revealed, the sperm from males reach an Amazon molly's eggs and fuse with them, injecting their genes. But the paternal and maternal genes don't organize themselves into a new genome. Instead, enzymes in the eggs shred the would-be father's DNA. All that an Amazon molly needs from a male is a trigger that starts her eggs turning into embryos.

And that's why the Amazon mollies make trouble for those who would draw sharp lines around life. One Amazon molly cannot reproduce. But two Amazon mollies cannot, either. In fact, the entire species of Amazon mollies is unable to create offspring on its own. The fish are sexual parasites, depending on other species for their reproduction. If life must be defined as a species that can self-reproduce, then these outwardly ordinary fish straddle its edge.

Of course, Amazon mollies are not entirely separated from more ordinary forms of life. They descend, after all, from mollies that display all the familiar hallmarks of life. The same is true for the other straddlers, the other half lives we can find around us today. Mitochondria descended from run-of-the-mill ocean bacteria that just so happened to get guzzled by our single-celled ancestors, entering 2 billion years of a twilight existence. Even viruses can often be traced back to rogue bits of parasitic DNA that started out in ordinary organisms.

But if we push back further, perhaps 4 billion years, all of life gives way to half life, and then to no life at all.

DATA NEEDED FOR A BLUEPRINT

David Deamer looked out across the crater and felt as if he were standing on an infant Earth. It had taken him days to get here, first flying from California to Alaska, and then over the Bering Sea to Russia's eastern fringe. In the city of Petropavlovsk-Kamchatsky, Deamer boarded an old army troop carrier with a team of American and Russian scientists, and they drove for five hours to the mouth of a canyon. The crew hiked into the canyon on a muddy trail that eventually ascended the slope of Mount Mutnovsky, a lively volcano. The year was 2004. Mutnovsky had last erupted in 2000.

At sixty-five, Deamer was Lincoln tall and Eisenhower bald. He clambered around looming boulders, past ash and frozen lava flows. The horizon crested with the peaks of neighboring volcanoes. After climbing 2,000 feet, Deamer and his fellow scientists reached the rim of Mutnovsky's crater. Nothing grew there on the expanse of black and gray rock. Steam roared out of the ground. Deamer put on a gas mask and descended into the maw. For the next few days the team of scientists surveyed the crater of the volcano and then its flanks. They collected samples of water and mud. And then Deamer began an experiment.

His lab bench was a field of boiling hot springs with the rotten-egg stink of hydrogen sulfide. For his test tube Deamer picked out a puddle the size of a modest pothole. The water, as acidic as vinegar, was loaded with whitish clay. At the center of the puddle, a column of boiling bubbles pushed through the slurry.

On his climb up the volcano, Deamer had brought a powder of life he had concocted in California. Its ingredients included the four nucleotides of RNA, as well as four amino acids, the building blocks of protein: alanine, aspartic acid, glycine, and valine. Deamer had finished the powder off with myristic acid, a component of coconut oil.

Deamer dipped a beaker into the scalding water and scooped up a liter. He sprinkled in his powder, and the water turned milky. Once it was well mixed, he leaned carefully over the puddle and poured out the solution.

He was doing something akin to what John Butler Burke had done a century before. To understand the nature of life, he was carrying out an experiment, putting lifeless chemicals into a container where they might take on some of the properties of living things. While Burke was a physicist with little understanding of the molecular basis of life, Deamer had four decades of modern biochemistry under his belt. But even with all that expertise, Deamer couldn't predict what would happen next on the volcano.

As soon as he emptied the beaker into the pond, a white froth appeared on its steaming surface. Nature had surprised him once more. Deamer bottled some of the froth and scraped some clay to take home, in the hopes of getting a little closer to understanding how life began 4 billion years ago—perhaps at a place like Mount Mutnovsky.

"It is mere rubbish thinking, at present, of origin of life; one might as well think of origin of matter," Charles Darwin wrote to his friend Joseph Hooker in 1863.

Charles was far more conservative than his grandfather Erasmus. He refused to speculate in public about how life might have arisen from lifeless matter. Writing *On the Origin of Species*, he alluded to the question only once. "Probably all the organic beings which have ever lived on this earth have descended from some one primordial form, into which life was first breathed," Darwin wrote.

Darwin would come to regret using that last word. "Breathed" was redolent of biblical creation. The only thing Darwin meant to convey was that living things must have arisen at some point in the distant past. How it happened, he couldn't say.

In another letter to Hooker, Darwin mused about how a "warm little pond" might serve as a flask for chemical reactions that produced simple organisms. He never shared that notion in public, let alone developed it into a full-fledged theory. But to his friends he confided how thrilled he'd be by the discovery that life arose from chemicals, "for it would be a discovery of transcendent importance." He would be just as thrilled if someone disproved it.

"But I shall not live to see all this," he predicted.

Darwin's reticence disappointed his disciples. Their hero had developed a theory that made it possible to tackle one of the greatest questions in science, only to stop in his tracks. "The chief defect of the Darwinian theory," Ernst Haeckel complained, "is that it throws no light on the origin of the primitive organism—probably a simple cell—from which all the others have descended. When Darwin assumes a special creative act for this first species, he is not consistent, and, I think, not quite sincere."

Haeckel and other followers of Darwin did not hesitate to take the leap. They marshaled evidence for how life may have begun. They wrote books, gave sensational lectures, and battled religious opponents who declared God alone could bring life into existence. But as they walked along life's edge, they found the path dangerously slippery. Huxley thought he had discovered the planet-spanning *Bathybius*, only to discover bad chemistry had led him astray. John Butler Burke may have been one of the first scientists to try to rerun the origin of life in a test tube. But within a few months of becoming a worldwide celebrity, he sank from view.

In hindsight it seems foolish to have even tried to trace the origin of life in an era when scientists knew so little about life itself. Huxley could talk about protoplasm, but only in terms that made it sound like a near-mystical jelly. When it came to heredity, no one in the nineteenth century—not even Darwin—could make sense of it. Huxley had been dead for five years when the very word *genetics* was coined in 1900. In the first few decades of the twentieth century, biologists finally laid spontaneous generation to rest. They began deciphering a few enzymes and tracking a few genes through generations of fruit flies.

In the Soviet Union, a biochemist named Alexander Oparin became convinced that these advances had finally made it possible to start thinking sensibly about the origin of life. At last science had safely put vitalism behind. "The numerous attempts to discover some specific 'vital energies' resident only in organisms invariably ended in total failure," he concluded.

To Oparin, it was hard to distinguish living things from the rest of the universe. Our bodies were made up of carbon, oxygen, and other elements that could be found in ocean waves, stratospheric clouds, and grains of sand. Our bodies used enzymes to make new molecules, but some of the same chemical reactions could take place outside a living thing. Living things could grow in complex patterns, but crystals could, too. The flowerlike crystals of ice that formed on windows in winter were evidence enough.

"In their delicacy, complexity, beauty and variety these 'ice flowers' may even look like tropical vegetation while all the time being nothing at all but water, the simplest compound we know," Oparin said. The reason that ice flowers were not in fact alive was that they lacked some of the other features required for life. "Life is not characterized by any special properties," Oparin concluded, "but by a definite, specific combination of these properties."

Looking at life this way made understanding its origins less daunting. The question of how life began was not all that different from how Earth began. By the 1920s astronomers already recognized that the solar system had started out as a disk of dust. Gravity caused the grains to draw together, to clump and crash until they formed planets. When Earth formed, it was a ball of molten rock. Over millions of years it cooled to form a hard crust. The atmosphere rained down an ocean. Oparin saw all of these transformations as a grand chemical experiment producing all sorts of new compounds that could then react with each other to create still more compounds, which gradually joined together all the properties required for life.

Oparin laid out some of his ideas in a small book in 1924. He wrote it in Russian, and only a few of his fellow Soviet scientists read it. But that disappointing reception did not cause Oparin to abandon his train of thought. Instead he ran experiments and read widely. He wove together new ideas from microbiology to chemistry, geology, and astronomy, seeing connections between the fields that narrow experts might have missed. In 1936, Oparin turned those new insights into a much longer book, *The Origin of Life*, which was translated into English, reaching a far bigger audience. He opened his readers' minds to a crucial realization: the planet on which life began was profoundly different from the planet on which we live today.

We breathe in air that is 21 percent oxygen. The oxygen molecules in the atmosphere steadily vanish because they react easily with other compounds. The planet's oxygen supply is replenished by plants, algae, and photosynthetic bacteria. Before life began, the atmosphere would have been almost oxygen-free. Oparin recognized that chemical reactions on such a world would operate in a profoundly different way than they do today. And he argued that some of these reactions produced the first building blocks of life.

Oparin speculated that steam from volcanoes could react with minerals to produce hydrocarbons. Those hydrocarbons could, in turn, go through other reactions to produce more complex compounds. The compounds started to clump together and began to grab molecules from their surroundings. They built more clumps like themselves and gradually turned into cell-based life as we know it.

No time machine could carry Oparin back to the young Earth to see if his own scenario was correct. Scientists would have to carry out experiments and gather clues from Earth and other planets to test their hypotheses and develop better ones.

"The road ahead of us is hard and long," Oparin warned, "but without doubt it leads to the ultimate knowledge of the nature of life."

Oparin was not the only scientist musing about the young Earth in the 1920s. J. B. S. Haldane published an essay of his own on the origin of life in 1929. Although they were unaware of each other, Haldane's and Oparin's thoughts ran along parallel tracks, back to the time when living things first emerged. "We may, I think, legitimately speculate on the origin of life on this planet," Haldane wrote.

Like Oparin, Haldane recognized that the differences between Earth now and at its birth would be crucial to those speculations.

He mused about ultraviolet light acting on water, carbon dioxide, and ammonia, producing sugars and amino acids that would accumulate in the ocean until it gained the consistency of what he called "hot dilute soup."

For all the similarities in their thinking, Oparin and Haldane emphasized different aspects of life. Oparin saw it fundamentally as a chemical problem. Look in the index of *The Origin of Life* and you will find plenty of entries about metabolism, such as *hydrolysis* and *oxidation.* But there's no *gene,* no *heredity.*

Haldane was first and foremost a geneticist, and for him the great question about the origin of life was how it started to copy its genetic information. He held that genes emerged early in the origin of life. Our genes may be swaddled today in deep layers of proteins and membranes within our cells. But the first genes must have been naked molecules building copies of themselves out of Haldane's hot dilute soup.

A generation after Haldane and Oparin first put forward their ideas, a graduate student at the University of Chicago heard about them for the first time. Sitting in a departmental seminar, Stanley Miller was intrigued but also puzzled: Why had no one successfully tested these ideas yet? Miller wasn't interested in doing the experiments himself; he considered experiments in general to be a messy waste of time. He preferred lofty theoretical science instead, and was planning to spend his time in graduate school pondering how stars made new elements.

Those plans fell through when his advisor left Chicago for a job in California. Desperate for a research project, Miller thought back to the origin of life. The more Miller thought about it, the less crazy an experiment to test Oparin's ideas seemed. He was not going to make radiobes, let alone full-blown life. He would merely test the proposition that the chemistry of the early Earth gave rise to organic molecules.

The seminar where Miller had learned about Oparin had been presented by a Nobel Prize–winning chemist named Harold Urey. Miller tracked Urey down in his office and proposed his plan. Urey replied that it was a bad idea for a graduate student, since it was likely to end in failure. He tried pushing Miller to other projects that were less ambitious but more reliable, like cataloging the chemicals in meteorites. But Miller wouldn't budge, and Urey eventually relented. He allowed Miller a year to tinker with the experiment in his lab. If Miller didn't make any progress after a year, he'd have to move on.

For the experiment, Miller and Urey set out to mimic the early Earth on a countertop. "We then designed a glass apparatus that contained a model ocean, an atmosphere, and a condenser to produce the rain," Miller later recalled.

Into this flask Miller added gases believed to be common on the early Earth: water vapor, methane, ammonia, and hydrogen. The energy for chemical reactions on the early Earth might have come from lightning, Miller speculated, so he inserted electrodes into the apparatus to deliver sparks. After a few initial trials and adjustments, Miller powered up the apparatus and let it run overnight.

The next day the solution had turned to a reddish muck. When Miller emptied the flask, he found that the muck now contained amino acids—the building blocks of proteins—along with a host of other carbon-bearing molecules.

Miller published his results in May 1953, at the tender age of twenty-three. "The reaction to the paper startled me," he later recalled. Like John Butler Burke before him, Miller was beset by a swarm of reporters. The news of his experiment was so sensational that Gallup conducted a poll to find out how many people thought it was possible to create life in a test tube. Only 9 percent said yes.

With that one experiment Miller created a new field of science that came to be known as prebiotic chemistry. Scientists created more amino acids and even some of the bases that today are parts of DNA and RNA. Haldane, who had helped seed the field with ideas as a young man, now looked on from old age at the new discoveries. He also found inspiration in the work of molecular biologists like Francis Crick, who were working out how life could store information in genes and then extract it.

Even in the 1960s, Haldane had fresh ideas to sow. Life, he came to believe, was the "indefinite replication of patterns of large molecules." The first patterns must have been much simpler than the ones that surround us today. The fact that some viruses used single-stranded RNA instead of double-stranded DNA got Haldane thinking that perhaps RNA evolved first.

In 1963, Haldane traveled to Florida to talk about his ideas at a conference attended by Oparin and other leading researchers on the origin of life. Haldane entitled his talk "Data Needed for a Blueprint of the First Organism." He envisioned a long-vanished form of life, a *Bathybius* for the modern era. It was a free-living microbe that stored its genes in RNA, not DNA. It could use its RNA genes as a guide to build proteins, which could then make new copies of its own genes. Just how few genes such an RNA-based life-form needed, Haldane couldn't say. "The initial organism may have consisted of one so-called 'gene' of RNA," he speculated.

The idea was potent—so potent, in fact, that it independently occurred to Crick and other scientists as well. But Crick, Haldane, and every other scientist who promoted RNA-based life could speak of it only in the haziest of terms. On the modern Earth, the only RNA-based life-forms are viruses, which need a host to reproduce them. On the early Earth, RNA-based life would have had to fend for itself.

David Deamer's journey to a Russian volcano began in 1975, over cucumber sandwiches on the side of an English road. He was having lunch with a British biophysicist named Alec Bangham, and the topic of conversation was membranes.

Life depends on genes for heredity and on proteins for its metabolism, but it also needs membranes to survive. They are the boundaries that keep life's chemistry bottled up and busy. Life, as far as we know, cannot exist as a boundless cloud of chemicals. But it wasn't until the 1950s that scientists like Bangham began pulling apart membranes and figuring out for the first time what they're made of.

One of the most common kinds of molecules in membranes is chains of carbon atoms called lipids. Some types of lipids are short and some are long; some have decorations of elements such as oxygen that alter their chemistry. But lipids all share a remarkable power of self-organization. One end of the lipid chain repels water molecules. The other attracts them. If loose lipids float in water, they spontaneously assemble into a two-layer film. The water-repelling ends tuck inside, while the water-loving ends face out. In the early 1960s, Bangham shook these films and found that they fell apart and then re-formed in three-dimensional shapes. At first they formed snakelike tubes. Then they pinched off into hollow spheres. These oily shells came to be known as liposomes.

Deamer, eight years younger than Bangham, had studied lipids in graduate school at Ohio State University, extracting them from egg yolks, spinach leaves, and rat livers. He traveled to California to become a postdoctoral researcher at Berkeley, where he learned how to freeze membranes and then crack them open to examine their inner structures. Deamer continued this line of work when he got a job at the University of California, Davis. At age thirty-six he arranged to spend a year working with Bangham in England.

The two scientists carried out a series of important new studies on lipids. They invented a syringe that could produce an abundance of liposomes of uniform size. Advances like these would turn liposomes into a medical tool. Drugmakers would later insert their compounds in liposomes to deliver them inside cells. When Covid-19 struck, vaccine makers slipped viral genes into liposomes, which could sneak them into our cells.

One day in 1975, Bangham and Deamer took a drive to London. When they stopped by the side of the road for lunch, Deamer mentioned hearing that Bangham had ideas about how life began. He was curious to hear what they were.

Bangham replied that life began with liposomes.

For Haldane and his intellectual descendants, genes above all other things make life special. For Oparin's followers, the great question about the origin of life is how metabolism arose. But life could not have arisen without boundaries as well, and Bangham's work with lipids gave him an idea for how the first primitive cells had formed. If lipids existed on the early Earth, they would have spontaneously turned into liposomes—ready-made containers for life's molecules. It would take far more time for the planet to produce primitive forms of DNA, RNA, and proteins. The one great shortcoming of Bangham's idea was that no one could say if lipids were indeed present before life began. And even if they were, no one knew if those primordial lipids had the right form to become hollow shells that could shelter life.

After Bangham and Deamer chatted about these profound matters, they finished their sandwiches and drove on to London.

"I thought, 'I'm going to go back to Davis to find which lipids can do this,'" Deamer later told me.

One of Deamer's graduate students, Will Hargreaves, volunteered to test out an assortment of lipids. He worked his way down from long lipids to short ones. Most lipids in living cells are twelve to eighteen carbon atoms long, but Hargreaves found that lipids with just ten carbons apiece could still make stable liposomes.

When Hargreaves finished his degree in 1980, Deamer was left wondering whether indeed the early Earth could have supplied these short lipids. Soon afterward he met a NASA scientist named Sherwood Chang who gave him a chance to find out. Chang was in possession of an extraordinary marble-sized rock, and he was willing to give Deamer a piece of it.

The rock had once been part of an asteroid that formed at the birth of the solar system 4.57 billion years ago. Another asteroid crashed into it, ejecting a meteor that then wandered the solar system until 1969, when it arrived in our cosmic neighborhood. Earth's gravitational field greedily drew the meteor in, and one morning the residents of an Australian town called Murchison looked up to see fireballs trailing smoke across the sky, followed by a clap of thunder. When people fanned out across the surrounding outback, they found hundreds of black stones.

NASA researchers got hold of some of these stones and discovered they were actually loosely joined mineral grains. Placed in water, they simply fell apart. Even more remarkable was what was inside the grains: amino acids, along with a host of other organic compounds. Life did not have to depend only on the chemistry taking place on our own planet for its ingredients, the Murchison meteorite showed. Many of its building blocks formed in space and then fell to Earth.

Chang gave Deamer a tiny sample of the Murchison meteorite. Back at Davis he treated it with chloroform and other chemicals to extract any lipids it might contain. He put the chloroform liquid on a slide to let it evaporate. It gave off a musty smell that gave him hope he had found something.

Once the chloroform had vanished, Deamer moistened the slide and peered through his microscope. He saw movement, organization. The water penetrated the dried extract, which swelled and grew into spheres. He had made liposomes. Deamer got out his camera and furiously photographed them. It was a moment worth memorializing, one over 4.5 billion years in the making.

The experiment suggested that lipids raining down from space might have spontaneously formed stable liposomes. But on their own the liposomes would be nothing but hollow shells. Deamer and his students began playing with mixtures of liposomes and organic molecules to see if they could fill the shells with the precursors of life. If they dried liposomes and DNA and then returned them to water, the liposomes re-formed with DNA inside them.

These experiments led Deamer to imagine a protocell with an enzyme inside that could build RNA molecules. But in order to build RNA, it would need a supply of bases. If bases were also produced on the early Earth, it might be able to pull them in. But this solution led to a problem of its own. Our cells pull compounds in from their surroundings through special channels encoded by our genes. Early protocells must have had a far simpler way to pull in their bases. Perhaps a molecule drifting past a protocell could get stuck on its membrane. Then it might slowly get drawn inside.

Deamer and his colleagues decided to build a model of a primordial membrane to see how it might work. They made sheets of lipids, into which they lodged proteins. They then added compounds to see if the proteins could shepherd them from one side of the sheets to the other.

In 1989, Deamer took a break from this work for a vacation in Oregon. On a long drive along the McKenzie River, he kept thinking about protocells and how they might pull in molecules. His mind wandered until he was daydreaming about streams of bases flowing into protocells through primordial channels. He would need a way to pull them through—perhaps an electric field. He pictured the base slowly wiggling through the channel, blocking the smaller charged atoms behind it like a slow-moving truck with a line of cars piling up behind it. It occurred to Deamer that this traffic jam would slow the current through the channel momentarily. He wondered what would happen if he and his students measured the channel's current as the base went through.

"Maybe we'd see a little blip," he later recalled.

What if, instead of a single base, Deamer tried snaking a piece of DNA through? Instead of a blip, would he see a series of blips? Each of the four bases in DNA has a different size and shape. Maybe the blips would look different. Maybe he could spell out the sequence in a piece of DNA by pulling it through a channel.

Suddenly, in the middle of the Cascade Mountains, Deamer realized that thinking about the origin of life had led him to something he never anticipated. He was thinking of a way to read DNA.

In 1989 the idea of quickly reading a piece of DNA was close to magic. The standard methods at the time were so slow that scientists could read only a few hundred bases each day. At that pace they'd need upwards of 100,000 years to sequence a single human genome. Some scientists were dreaming of ways to speed up the process, and now Deamer had become one of the dreamers. He imagined DNA shooting through a channel, singing out its sequence in an aria of electricity.

When Deamer finished his Oregon drive in 1989, he took out a red pen and drew his vision in a notebook. He sketched DNA slipping through a channel. He drew an imaginary graph showing the blips of voltage he imagined each base would create. "The channel must be of the dimensions of DNA in cross-section," Deamer wrote.

Deamer enlisted other scientists to help him make the idea real. For starters, they would have to find a channel of the right size and shape to create a DNA traffic jam. In 1993, Deamer learned about one that might do the trick, a channel made by bacteria called hemolysin. He traveled to the laboratory of a hemolysin expert named John Kasianowicz at the National Institute of Standards and Technology in Maryland, bringing with him strands of RNA to thread through a molecular needle.

Together, Deamer and Kasianowicz created a lipid membrane stretched across a circular opening. In the middle of the membrane, they inserted a single hemolysin channel. When they switched on an electric field, they could drag RNA into the hole. And they could see a series of blips. The number of blips matched the number of bases on the strands.

That success was enough to publish a paper in 1996. But it was still a long way from a DNA reader. They had yet to figure out how to tell all four bases apart. It was as if Deamer and Kasianowicz were looking at blacked-out sentences in a redacted government document. They could count the number of letters in the sentences but had no idea of the words the letters spelled out.

One of Deamer's former students, Mark Akeson, came back to California to take over the project. His goal was to pull the mask away from the letters. Akeson and his colleagues tuned their electronics to detect even subtler changes in the current while making them less sensitive to distracting noise. They took advantage of the fact that two of DNA's four bases, adenine and guanine, are much bigger than cytosine and thymine. Akeson and his colleagues proved that the big bases led to big drops in the current, and the small bases led to small ones.

Deamer could not yet hear the language of genes clearly. But now, at least, he could tell its vowels from its consonants.

I first met David Deamer in 1995. I traveled to Santa Cruz, where he had moved after marrying Ólöf Einarsdóttir, a professor at the University of California campus at the north end of the city.

Deamer had traded Davis's landscapes of flat farmland for the brooding beauty of the coast, a place where elephant seals lounged on the beaches, watched over by pines and redwoods on the hillsides. On my first night in Santa Cruz, I wandered downtown. The Loma Prieta earthquake of 1989 had left its mark six years earlier. I passed silent abandoned buildings, following the stark gashes in the dark, deserted streets. In the morning I found my way to Deamer's lab.

"Do you want to smell outer space?" Deamer asked. He offered me a sample of Murchison lipids to sniff. It reminded me of an attic. "Do you want to hear insulin?" he asked. A few years before, Deamer converted the sequence of genes into musical notation: adenine became A, guanine became G, cytosine became C, and thymine—without a T in the scale—became E. He began to hum a gene to me, which sounded vaguely like a song.

Deamer was fifty-six at the time. A decade had passed since he made liposomes from a meteorite, and in the intervening years he had developed an elaborate scenario for the origin of life, drawing on both his own work and that of other scientists. The idea that life started out based on RNA, originally conceived by Haldane and others in the 1960s, had gained a lot of favor over the years. RNA had proven to be exquisitely versatile—perhaps versatile enough to sustain life on the early Earth. At the University of Colorado, for example, a biochemist named Thomas Cech discovered a remarkable RNA molecule in a freshwater protozoan called *Tetrahymena*. This molecular strand could spontaneously bend around and cut out a piece of itself, like a self-acting enzyme. Soon researchers were finding other RNA molecules that can behave like enzymes—what came to be known as ribozymes.

Ribozymes revealed that RNA can do two things at once: they can store genetic information like DNA, and they can also carry out enzymatic reactions like proteins. In 1986 a Harvard biochemist named Walter Gilbert used their discovery to update the hypotheses of Haldane and others about the origin of life. He called his theory "the RNA World."

Gilbert proposed that life initially used RNA alone, long before DNA and proteins even existed. An RNA-based form of life might carry a set of RNA molecules, each adapted for certain jobs. Some might carry genetic information, while others might grab compounds to build new RNA molecules. RNA-based life could evolve because it would make mistakes as it made new copies of its genes.

Eventually, Gilbert proposed, RNA-based life evolved proteins and DNA. RNA molecules might have gained the ability to link amino acids together to make very short proteins. These new molecules may have been able to help the cells survive, and as the proteins got longer, they may have outperformed the RNA molecules. RNA genes may have evolved into the double-stranded form of DNA, which proved a more stable way to encode genes.

Gilbert followed in Haldane's gene-centered tradition. He focused entirely on the evolution of RNA molecules without giving any attention to how they would be housed in cells. Using his liposomes, Deamer pursued that unaddressed question.

He hypothesized that primitive cells might have formed from the lipids delivered by meteorites. Some of these meteorites might have landed on the newly forming volcanoes that rose above the ocean. The lipids were washed into ponds and hot springs—along with an assortment of other potential building blocks for proteins and RNA. Periodically the water would evaporate, leaving a kind of primordial bathtub ring behind, which was later submerged once more in rain or floods.

Working with Ajoy Chakrabarti, a postdoctoral researcher in his lab, Deamer re-created this ancient chemistry for me. He opened a jar of egg yolk lipids and added some to the water in a test tube. The tube turned cloudy as it filled with microscopic bubbles.

Deamer then turned to a second test tube, adding dried white threads of DNA from salmon sperm as if he were a chef sprinkling saffron into a dish. (Salmon sperm DNA is cheap and easy to order from a biological supply company; it stood in well enough for RNA.) The DNA threads turned gooey. Deamer spiked the solution with a fluorescent stain. He then combined the lipids and the DNA on some slides.

"Why don't we get the hot plate going?" he said to Chakrabarti.

Chakrabarti switched it on and put the slides on its surface.

"That's our tide pool," Deamer said.

In a primordial pool, lipids might have formed into liposomes drifting around in the water. But as the sun beat down, the water would disappear and the liposomes would get crowded together. When they touched, they fused. As more water evaporated, they turned from bubbles to sheets, sandwiching other molecules between their layers.

The same thing was happening on the slide. After a few minutes Deamer removed it from the hot plate. The DNA and lipids had dried to a thin film. Now Deamer refilled his miniature tide pool by adding back a few drops of water. He put the moistened slide under a fluorescent microscope, and Chakrabarti turned out the lights.

Through the eyepiece I saw lipids squirting out from the dried film into the surrounding water. At first they writhed like snakes, and gradually they swelled into bubbles. Some of the bubbles were dim, but others glowed with the intense fluorescent green dye, letting me know that they had swallowed up DNA.

This exercise was a far cry from a proof of how life began. Deamer just wanted to show one step in the scenario he and like-minded researchers favored. At the time, they were under heavy fire from RNA-World skeptics. No one could yet say how RNA molecules could come together from simple building blocks to begin with. As for where life might have begun, a number of scientists were looking away from the volcanic ponds that Deamer favored. They looked instead to the bottom of the ocean.

In the 1970s oceanographers investigated mid-ocean ridges, the seams between continental plates that run from pole to pole, where magma rises up from deep within the earth and adds new margins to the seafloor. The researchers were surprised to discover huge black chimneys sitting on the ridges, spewing dark smoke. It turned out the chimneys were deep-sea versions of hot springs. Ocean water was making its way down through the fissures in the ridges, where it heated up and reacted with the surrounding minerals. When it rose back to the seafloor, it brought a heavy load of subterranean compounds. Hitting the cold seawater, the minerals in the fluid suddenly underwent chemical reactions and formed hollow piles of rock on the seabed.

On closer inspection, scientists discovered that these vents harbored life—ecosystems unlike any others on Earth. Microbes harvested energy from the chemicals spewed from the vents. They became food for larger organisms. Blind shrimp crawled the flanks of chimneys. Tube worms grew like bamboo forests. More than 4 billion years ago, when the earth cooled from a molten ball and developed a crust, the early ocean would have contained many such vents. The heat and exotic chemistry present in them might have fueled the rise of genes, metabolism, and cells.

Deamer was not having it. The prebiotic manna from heaven—organic compounds falling from space—would get diluted in the ocean's great expanse before it could reach vents on the seafloor. Liposomes that formed in the ocean would get torn apart by its salty chemistry.

Still, Deamer had a lot of work before him. If life started in surface pools, it would need some way to get an energy supply. Today, algae and bacteria in ponds can harness sunlight, but they use a complex network of proteins to do the job. A protocell could not have relied on such a sophisticated natural solar panel. But

Deamer wondered if simple solar panels might already be floating around them. The Murchison meteorite contained molecules called polycyclic aromatic hydrocarbons, or PAHs for short. When light shines on a PAH, it can give off an electron.

Perhaps, Deamer speculated, PAHs from meteorites could insert themselves in liposomes. When sunlight hit them, the PAHs would release electrons that they could use. They could generate the power protocells needed to carry out their chemistry.

Nobody could say whether this scenario would work or not, because nobody had ever tried mixing PAHs with liposomes before. So Deamer and his students tried.

"We'd like to make them capture energy in a useful form," Deamer told me. "Nobody's particularly impressed yet."

Four years later, in 1999, Deamer met a Russian volcanologist named Vladimir Kompanichenko at a conference on the origin of life. When Kompanichenko learned of Deamer's obsession with primordial pools, he invited him to come to Kamchatka. It would be the closest thing to time travel Deamer would ever experience. The peninsula of Kamchatka was packed with active volcanoes, and the conditions were so harsh that little could survive there. If he made the trip, Deamer would be able to study crater lakes, hot springs, ponds, and all manner of other bodies of water. Rather than imagine the chemistry of the early Earth, he could look at it up close.

Deamer took Kompanichenko up on the offer. He organized a team of scientists to travel to Kamchatka in 2001. They rode a military helicopter from volcano to volcano, as brown bears scurried away on the tundra below. One volcano lake was turquoise blue, while another was topped with petroleum—not from an oil spill but from the swift breakdown of plant matter that blew into the water. Ordinarily it takes hundreds of millions of years for plant matter to turn to petroleum. In this strange place it took only centuries.

On the flanks of the volcanoes Deamer scooped water from steam-blasting fumaroles and inspected hot springs bordered by bathtub-like rings—exactly the wetting and drying cycles he hoped to find in nature. The ponds contained different combinations of minerals, reached different temperatures, and varied in many other ways. There was so much for Deamer to take in that he knew he would have to return. And it was on his second trip, in 2004, that he brought his powder of life.

For thirty years Deamer had been studying the origin of life in the tradition of Stanley Miller, working in the confines of a laboratory. He carried out experiments in glass tubes with pure ingredients and precisely controlled temperatures. Those controls allowed him to know whether his results were significant or not. But they also left him wondering if the processes he studied in his lab would work in the rough-and-tumble world where life has to survive.

And as soon as he poured his powder into the Mutnovsky puddle and the froth appeared, Deamer knew something strange had happened. The froth was made up of lipids that had organized into membranes. But he had to get back to Santa Cruz to figure out exactly what he had witnessed. He and his colleagues discovered that many of the compounds in his powder had gotten stuck to the clay particles in the water. But the lipids had captured others. The lipids did not immediately turn into bubbles as they might in Deamer's laboratory. Iron and aluminum in the water had reacted with the lipids and turned them to floating curds.

Deamer did not make life from scratch on Mount Mutnovsky, but the experience had a profound influence on his thinking. The ponds and hot springs on the volcano all had high temperatures and low pHs, but they were also different in many ways. Some ponds were laced with clay or aluminum that might block the development of life, while others might be more favorable. Deamer began surveying the diversity of hydrothermal springs in other parts of the world. Sometimes he visited himself, and sometimes he arranged for colleagues and students to do the fieldwork. They went to Yellowstone, Hawaii, and Iceland. On a trip to New Zealand, his colleague Bruce Damer brought an aluminum block loaded with test tubes. Each test tube had a dried film of RNA and other chemicals. Damer pushed the block into the mud and filled it periodically with water from the spring. They succeeded in producing liposomes containing small molecules of RNA.

These trips were expensive, demanding, and relatively brief. To continue the research back at home in Santa Cruz, Deamer built an artificial volcanic pond. "I'm mimicking what I saw on Mount Mutnovsky," he told me.

Deamer constructed a clear plastic box the size of a suitcase. He sealed it so that he could flood its interior with carbon dioxide, giving it an atmosphere more like the one that existed on Earth 4 billion years ago. Inside the box, Deamer installed a metal disk with holes around the edge into which he could slot two dozen tubes. Each tube could mimic a pond in Kamchatka, with hot acidic water laced with various chemicals like the ones he sampled on Mount Mutnovksy. He created his own cycle of drying and wetting. The disk slowly rotated, so that each tube passed under a tube blasting carbon dioxide for half an hour twice a day, evaporating its water and leaving behind bathtub rings of chemicals. As the disk rotated more, the dried tube moved underneath another tube that delivered a splash of water.

Deamer and his colleagues filled the tubes with lipids and bases, the building blocks of RNA and DNA. After the tubes went through hours of wetting and drying, they found liposomes with bases trapped inside them. And in a small fraction of these shells they found something even more remarkable: the bases had joined together. Some of the new molecules were up to a hundred nucleotides long. "We've made an RNA-like molecule," Deamer said.

In our own cells, bases form bonds only with the help of highly evolved enzymes. Deamer and his colleagues had sidestepped this requirement by using the peculiar chemistry of a primordial pond. When the liposomes dried, they fused and flattened into sheets. These thin layers became liquid crystals where the bases no longer bounced around in endless agitation. Instead, they fell into an orderly arrangement, in which they were more likely to bond together. When water returned to the tubes, the layers swelled and budded off as bubbles, taking the RNA-like molecules away with them. With each round of wetting and drying, the molecules got longer.

In Walter Gilbert's RNA World, the first living things needed a ribozyme to build RNA molecules. Now Deamer's experiments suggested something even more radical: no ribozyme was required because lipids could do the work of building RNA on their own. Before there was an RNA World, his work suggested, there might have been a Lipid World.

I visited David Deamer for a second time in the fall of 2019. Twenty-six years had passed since my first trip to Santa Cruz. I had become a gray-haired father, while Deamer had just celebrated his eightieth birthday. I had flown to San Francisco for work, and Deamer insisted on picking me up at my hotel and chauffeuring me down to Santa Cruz for an afternoon. He was in good health, I could see; he credited it to biochemistry. Experiments he had carried out in the 1970s had convinced him of the benefits of antioxidants, and so he started taking supplements. "And here I am, still going strong," he said.

Still, Deamer asked me to stay quiet so he could concentrate on getting us safely out of the city and onto the freeway. Once we reached the pine groves and the coastal cliffs, he relaxed. He began humming a song of DNA.

I asked Deamer what, after all this time, he thought life was. He still didn't have a good answer, he admitted. "We will know when the molecular systems we've assembled happen to have certain properties of life," he replied. Deamer then rattled off some of those properties. Was he defining life, I asked, or just characterizing life as we know it? Did life *have* to be based on chain-shaped molecules like DNA and proteins?

"I'm stuck in my little box," Deamer admitted. "I can't imagine anything other than a nucleic acid and proteins that can do what they do. You ask me how I think about all this. I like to do experiments. I like to watch things happen. I just think, 'What's the next simple thing I can do?'"

When we got to Santa Cruz, I could see that the damage from the earthquake had healed over since my last visit. But other fissures had opened up since then, ones that would be harder to fix. Wealthy tech workers priced out of Silicon Valley had streamed over the mountains, offering a million dollars for a petite bungalow. Near the town bus station I watched a woman slowly wander barefoot, mimicking a cigarette to passersby as a silent request.

Deamer did not take me to the redwood grove where his

university laboratory had once been. Instead, we headed to a warehouse-like building at the edge of town near the railroad tracks. The year before, he had launched a company in an incubator called Startup Sandbox. It housed start-ups developing bone grafts, cancer tests, and smart gardens. Deamer was three times older than most of the people there.

We settled into his second-floor office. It had the hollow feel of a place barely moved into. A framed photograph of a shooting star hung on a wall. A science fiction novel by Stanislaw Lem sat alone on a shelf. From under a table Deamer pulled out his artificial pond to show me how it worked.

"There's not much to it," he said, "but it's the only one in the world."

I, in turn, had something to show Deamer. I pulled up a picture on my phone that a biologist I knew had recently sent me. It showed a metal block the size of an eight-hole harmonica. Next to it was an opened box, labeled *MinION*.

"My new toy arrived," my friend texted. "$1000 sequencer. Less than your iPhone! I can't decide whether to be excited or horrified."

I scrolled down to the next message, which my friend sent me a few weeks later. He wanted to find out what sort of microbes grow on paintings, so he pried a speck of paint from an old artwork, extracted genetic material from it, and then placed a DNA-loaded drop into his MinION sequencer. He sent me a video of the MinION hooked up to his laptop, which was busy reading the DNA sequences. In five hours, the MinION had read 42 million base pairs.

"Ta da!" my friend texted. "Hard to believe this will feel old-fashioned in my lifetime."

"Oh, look at that!" Deamer said in soft delight. I wasn't surprised that the video would make him happy. The machine my friend was using had its origins in Deamer's dream three decades earlier.

In 2007 a company called Oxford Nanopore Technologies licensed the patent Deamer and his colleagues had filed for their concept of a DNA sequencer. In the years that followed, Deamer and other scientists found ways to improve the design. Better channels came to light in other bacteria. Oxford Nanopore figured out how to fit many channels on a single membrane so that they could sequence many copies of DNA at once. They also started spending time in court. As the technology grew more promising, other DNA-sequencing companies started challenging the patents. "We're continually being sued," Deamer told me.

Oxford Nanopore began selling their first DNA reader in 2015. Compared to other technologies, it was tiny, easy, and cheap. Scientists began using it to read DNA that would have otherwise gone unread. During the 2015 West Africa Ebola outbreak, it took only a day for scientists to read the genes of viruses after they were extracted from patients. In the forests of Uganda, wildlife biologists rapidly identified new insect species. In 2016, NASA sent the MinION to the International Space Station, where the astronaut Kathleen Rubins carried out the first DNA sequencing in space. Someday, Deamer hoped, a nanopore sequencer might discover genes on another planet.

Deamer's ideas were becoming real in another way: younger scientists were building more elaborate protocells to explore the RNA World. A biologist named Kate Adamala came up with her own recipe for lipid bubbles, into which she slipped RNA molecules. She created protocells that grew and split in two. She created protocells that gave off a flash of light when they detected a certain chemical. She made protocells that could talk to each other. None of Adamala's protocells could do all of these things at once, though. She dealt each protocell a different hand of RNA. But collectively Adamala's creations offered glimpses at what life might have been like before life as we know it—if we are willing to accept something without DNA as alive.

In his own lab Deamer worked with his students to uncover more of the steps by which loose lipids and nucleic acids might have assembled into those first protocells. In 2008 they discovered that liposomes that went through cycles of wetting and drying could produce RNA molecules up to one hundred bases long. But skeptics observed that these molecules were much shorter than any RNA virus's genome. It was hard to see how such brief genetic instructions could launch life on its journey. So Deamer and his team tried to make bigger ones.

Not long before my visit, they began using a new tool to look at what they created. The device, called an atomic force microscope, taps a minuscule metal finger over molecules, mapping each of their atoms. Deamer showed me one of the maps. It was a biochemical Jackson Pollock: a field of strings, tangles, loops.

"If we're right, those are the longest strands ever made in the history of this research," Deamer said. "If you're going to make a ribozyme, it's got to be long enough to fold. We have ribozymes to spare here in terms of length."

The tangles gave Deamer more evidence for his vision of life's beginnings. After the earth formed, volcanoes rose above the ocean, and rain fell along their flanks. Ponds filled, and heated groundwater rose up through geysers and bubbling hot springs. Asteroids, meteorites, and dust fell from the sky, delivering trillions of tons of organic compounds. The volcanoes acted as chemical reactors, too, supplying their own compounds. When lipids reached bodies of water, they sometimes formed into bubbles, enveloping compounds and then delivering them to drying bathtub rings. RNA molecules grew in their liquid crystals, and when water returned, the dried layers became trillions of liposomes carrying new molecules.

Many of those bubbles ripped apart, but some remained stable. The RNA they carried inside acted as a brace, holding them together from within. These stabilized bubbles were more likely to survive long enough to get into the next bathtub rings, and their RNA was more likely to get into the next generation of bubbles. Deamer and his colleagues have found that a single strand of DNA can act as a template for a corresponding strand in these liquid crystals. On the early Earth, RNA molecules might have started getting copied in the bathtub rings, long before enzymes took over the job.

Over time, these networks of RNA added new molecules, and the molecules got longer. They took on new roles inside the liposomes. Some poked through the membranes to serve as primitive channels. Some snagged bases, speeding up the growth of new RNA molecules. The liposomes liberated themselves from their liquid crystal nursery and began to divide on their own. They may have powered their growth by trapping sunlight in pigments from meteorites.

In Deamer's telling, these protocells were the first truly living things. They were fragile organisms, to be sure. But without any competition, they could thrive. They evolved the ability to bring together amino acids, forming short chains and then longer ones that folded into true proteins. These proteins were stronger and more chemically versatile. Single-stranded RNA also evolved into double-stranded DNA, which proved to be a more stable way to store genetic information. In time, the new DNA-based organisms drove RNA-based life extinct.

In recent years paleontologists have been pushing back the fossil record of life on Earth, and some of the oldest evidence they've found comes from rocks in Australia dating back 3.5 billion years. They contain thick layers that may have been formed by microbial mats growing in volcanic ponds—precisely the places Deamer predicted that early life would thrive.

If you went back in time to the dawn of life, you might have observed puffy cushions of microbes lining bubbling springs on the flanks of volcanic islands. The islands were otherwise bare black rock sprinkled across a green ocean under an orange sky. Clouds sometimes rolled overhead, and rain washed down the islands. The streams moved microbes from pond to pond. In the ponds where microbes already resided, the newcomers mixed their genes with the old. The rain clouds traveled out to sea, leaving the islands to bake. The ponds dried out and winds picked up their dust, carrying microbial spores for miles. As they flew and swam downhill, the microbes reached salty estuaries. As they adapted to these new environments, they became ready to spread into the ocean. Once they reached the sea, the whole planet became alive.

"I'd give myself a hundred million years for something to happen," Deamer said.

I thought back to Darwin doubting he would live to see the origin of life settled. Here I was, nearly 150 years later, listening to a scientist talk about his life's work, dedicated to that mystery. I wondered what Deamer would live to see in the years he had left. Would the story he told grow stronger? Or would he be remembered as the John Butler Burke of his time?

He still had many opponents. One of the fiercest was a scientist named Michael Russell, who was also eighty years old. Russell's path to the origin of life was lined not with lipids but with minerals. He traveled to Pacific islands and Irish mines searching for seams of silver and fool's gold. In his journeys he recognized that some of those minerals had originally been produced around hydrothermal vents. These were not the superheated black smokers on the mid-ocean ridges, however. In other parts of the ocean, a different kind of chemistry took place.

In these places the seafloor is lined with olivine, a rock rich in magnesium and iron. Water flowing into the cracks reacts with the olivine, releasing hydrogen and heat. The rocks absorb the warmth and in turn boil the water, which shoots back up to the seafloor, bringing with it minerals, methane, and a host of other compounds. Many of the positively charged hydrogen atoms get combined into these compounds, which changes the fluid's pH. It goes from acidic to alkaline. When this hot fluid comes out of the seafloor and hits the cold, acidic bottom water, it dumps minerals that pile up into giant hollow chambers topped by towers that can reach two hundred feet high.

To Russell, these chambers seemed like the perfect place where life could arise. They became extraordinary chemical reactors, thanks to the difference between the water within and outside the chamber walls. The high pH of the alkaline water inside the chamber attracted the hydrogen atoms from the acidic seawater outside. The hydrogen atoms would have to make their way through microscopic channels in the walls. To Russell, their flow was a striking parallel to the way hydrogen atoms flow through channels in cell membranes—a flow our cells exploit to capture energy. In fact, Russell didn't think this was a coincidence. Our metabolism was built on the chemistry of the chambers.

The hydrogen atoms streaming into the walls of the alkaline vents could have powered chemical reactions, creating new compounds that could go through reactions of their own. Over time, the chambers created many of the ingredients necessary for life.

Russell speculated that pockets in the minerals might serve as cells before cells existed. In these rock-lined chambers, a primitive metabolism could grow. Eventually that metabolism became able to support full-blown life.

By the early 2000s alkaline vents and volcanic ponds were the two leading scenarios for how life began. Both could not be right. In a 2017 cover story for *Scientific American*, Deamer and his coauthors made their case for life forming on the planet's surface, complete with elaborate diagrams of life starting high on volcanoes and flowing downhill to the sea. Their scenario, they argued, found much more support from experiments than Russell's alkaline vent theory.

Russell struck back with barbs. In the work of scientists like Deamer, he saw that vitalism was alive and well. Experiments that generated lifelike molecules in mimic ponds were "utterly irrelevant and misleading." The idea that cells could have started out as liposomes bobbing in water was, Russell declared, "fundamentally flawed."

Only alkaline vents, Russell argued, offered the right flow of energy for producing the specific reactions that living things use today. Heating up a volcanic pond and bathing it in sunlight would do nothing but create a lot of competing reactions that would not add up to any complexity. The idea was as ridiculous as Dr. Frankenstein using jolts of electricity to bring dead body parts back to life.

"The Frankenstein idea is just false, whether it be cast in chemistry or in corpses," Russell declared.

When I brought up Russell's alkaline vent theory, Deamer cataloged the many problems he saw in it. One major flaw was that the walls of vents were too thick to generate the kind of energy Russell needed. Think about trying to generate electricity with a water wheel, Deamer said. If you put the water wheel directly under a waterfall, you can capture the energy of water falling a great vertical distance while flowing only a short distance downstream. "But if you take the waterfall over a kilometer, you cannot run a generator, because you just don't have that energy," Deamer said.

On the day of my visit, Deamer was running a new experiment. The reason he had brought me to Startup Sandbox was because he had recently launched a company called UpRNA. A Santa Cruz graduate student named Gabe Mednick had signed on as his sole employee. His microscopic company had one mission: to create another biotechnology based on the peculiar chemistry at the origin of life.

In August 2018, for the first time in its history, the U.S. Food and Drug Administration approved a drug made of RNA. A company called Alnylam Pharmaceuticals created it as a way to treat a disease called transthyretin amyloidosis. The disease is brought about by a mutant gene that creates a defective protein. Over the years, the damage from these faulty proteins causes people to waste away, to struggle to walk, to suffer seizures, and to die of heart attacks.

To create their drug, Alnylam created liposomes, using the techniques Deamer had helped pioneer forty years before. Into these oily bubbles they placed custom-made RNA molecules. The liposomes slipped inside cells, where they released the artificial RNA. These molecules then grabbed onto the messenger RNA for the faulty gene, preventing the cell from using it to make the faulty proteins.

Alnylam's success raised the prospect of using RNA to fight high cholesterol, cancer, and other disorders. But there was a catch: it costs a lot of money to make an RNA-based drug. Alnylam mimicked nature, using enzymes to read an artificial gene and then building the RNA molecule one base at a time. When Alnylam got approval for their drug, they put the price for a year's supply at $450,000.

Deamer suspected he could make custom RNA molecules for a lot less money. Instead of mimicking life as we know it today, he would mimic life in the RNA World. Deamer and Mednick were putting genes of DNA into his artificial pond and then trying to build matching RNA molecules by drying and wetting the tubes. For their first try, they were going to make an RNA molecule that would switch off a gene for a glowing protein. If he added the RNA to a dish of luminous cells, it would grab the messenger RNA for the protein and the cells would go dark. And if they could reach that milestone, they would start work on RNA molecules that could block disease-causing proteins.

Deamer had no idea if it would succeed. But all the lessons he had learned about life as it first existed gave him confidence that it might. "Everything I do," he said, "is based on knowing that life began."

NO OBVIOUS BUSHES

The February sunshine was almost too much to bear. I emerged out of a Lyft in front of the badging office at the NASA Jet Propulsion Laboratory in Pasadena, California, having just spent several months in chilly New England under a sky that had been mostly cloudy, dark, or both. A JPL scientist named Laurie Barge came into the office to meet me and led me inside the facility. Along the way, she slipped on a pair of Michael Kors sunglasses, which hid her eyes behind obsidian walls. As we walked across a palm-lined courtyard, I squinted like a miner just rescued from a cave-in.

I had come to JPL to talk to Barge about her work on astrobiology. "It's, essentially, how does life start and how do we find it?" she said as we settled down in the shade with coffee. It was the right kind of science for someone like Barge, who was full of big questions as a teenager. "I wanted to know why we're here," she said. "I wanted to know: Where did the sun come from? Why is there a universe? Why is there Earth? Why is Earth the way it is? Is life specific to Earth?"

It was hard to imagine someone like Barge anywhere else but JPL. It was one of the most important places on Earth for investigating the possibility that we are not alone. My visit with Barge was my first chance to see JPL for myself, and I will admit feeling like a pilgrim finally paying a visit to a sacred site.

I belonged to the generation born in the 1960s, when life on Earth suddenly extended its tendrils past the bacteria-laden stratosphere and into space. We sat cross-legged on carpets in front of bulging glass television screens, watching fuzzy images of two-legged mammals, encased in little chambers of Earth's atmosphere, walk on the moon. In the movies and TV shows we watched, humans traveled across the galaxy and met an endless procession of other life-forms that had an odd resemblance to the two-legged mammals who go to Hollywood casting calls. It seemed like the interplanetary age had begun.

Astronauts only got as far as the moon, though, and they never stayed long before coming home. By the 1970s their ambitions had contracted down to low Earth orbit. Their cramped space stations flew overhead, close enough to leave a glint on the night sky. Instead of humans, it was machines that explored the other planets. And many of those machines came from JPL.

It was at JPL that engineers built the first spacecraft to visit another planet—*Mariner 2*, which flew past Venus in 1962, and *Mariner 4*, which paid humanity's first visit to Mars three years later. It looked down at the red planet and snapped pictures, and when the pixels arrived at JPL, scientists saw a cratered desert.

The scientists at JPL included geologists who investigated how the planets had formed from clouds of pebbles. There were atmospheric scientists pondering swirls of carbon dioxide and sulfur dioxide. And JPL had biologists on staff, too. They did not study life as we know it, so much as contemplate life as it might be. In 1960 the microbiologist Joshua Lederberg had given this new field a new name: exobiology. One scientist scoffed that exobiologists were actually ex-biologists. The JPL exobiologists brushed away the mockery and worked on ways to detect extraterrestrial life.

The exobiologists at JPL could be roughly split into two camps. Some of them thought the best way to search for life was to search for it up close. They wanted to send probes to other planets, where they could run experiments. In one project, a group of exobiologists built a growth chamber into which they could scoop dirt on another planet and watch organisms metabolize carbon dioxide or some other gas.

Other exobiologists at JPL thought they'd have more luck by looking for life from a distance. The most prominent voice for a distant search was James Lovelock, a British-trained scientist who worked at JPL in the 1960s. To Lovelock, the idea of putting a growth chamber on Mars was absurdly provincial. We ought not limit our search for life to life that played by exactly the same rules as, say, bacteria that live in the ground on Earth. To Lovelock, the crucial fact of life was that it had the power to push chemistry away from equilibrium—not just within its own cells but across our entire planet. It flooded oxygen into the atmosphere. It eroded rocks, sending minerals into the ocean. At JPL, Lovelock tinkered with devices that might be able to see the signature of life in atmospheres on distant planets.

At first Venus and Mars seemed like the best places to start looking. They were practically neighbors, for one thing, and they both had a solid crust—as opposed to the gas giants farther out in the solar system. When *Mariner 2* visited Venus, however, it proved to have an atmosphere that trapped enough heat from the sun to melt lead. Exobiologists struck it off their very short list. As for Mars, the pictures that *Mariner 4* sent back weren't quite so bleak. The Martian deserts were cold and cratered. But even on Earth, life wasn't limited to lush rain forests. Mars might overlap with the harshest habitats on our own planet.

"The fact is that nothing we have learned about Mars—in contrast to Venus—excludes it as a possible abode of life," Norman Horowitz, the head of JPL's bioscience division, said in 1966. "We can say although the situation is not brimming with hope, neither is it hopeless."

Two years later NASA moved forward with the Viking mission: a project to put a pair of spacecraft in orbit around Mars. Each would drop a probe to the planet's surface for a closer look. The Viking rockets finally left Earth in the summer of 1975.

At the time, I was nine. It took the spacecraft almost a year to reach Mars, which is practically a geological epoch for a fourth grader. I waited and waited, hoping that the probes would discover Martians. They didn't have to be Hollywood extras. I'd have been fine with Martian snakes or skunks. Even a shrub or some bacteria would do, at least for the time being.

While the Viking probes flew through space, my family moved from a suburban house to a little farm in the country. Life now seemed to constantly demand my attention. Snapping turtles lurked in ponds; swallows darted in and out of the barn all day; cicadas whirred in the trees. I think now that this experience fooled me into thinking that life is hard to miss. I came to believe the history of space exploration would be written in short, quick headlines. MAN WALKS ON THE MOON. LIFE ON MARS.

In July 1976, *Viking 1* landed and sent its first pictures up to the orbiter, which relayed them in turn millions of miles to Earth, where JPL engineers deciphered them. We saw the first pictures on the evening news: gray rocks on a gray background. It was exhilarating and underwhelming at once. If I lay down on my driveway and looked out at the bits of gravel, I could have seen the same thing.

JPL held a live press conference where members of the *Viking 1* team watched the first images arrive. The astronomer Carl

Sagan stared at a nearby monitor, trying to make sense of them. In the months before the launch of *Viking 1*, he had mused that Mars might be home to multicellular organisms that would be easy to photograph. Now he strained to see anything at all.

"There certainly are no features, to the best of my knowledge, in these pictures which have to be due to life," he said, his eyes still locked on the monitor. "No obvious bushes, trees, or anybody else."

The following day, *Viking 1*'s first color pictures arrived. They painted the land red and the sky pink. Looking out for miles across the landscape, we saw no bushes, no trees, nobody else.

Later *Viking 1* put a shovel in the ground. It began analyzing the Martian dirt for signs of life. At the time, I was too young to understand what exactly they were doing to it; as far as I could tell, it seemed like they would cook it. Life being life, I assumed the tests would provide a clear yes or no.

The first tests were promising, but Horowitz warned the *New York Times* not to read too much into them. "We have not discovered life on Mars—*not*," he said. And when *Viking 1* looked in two samples of soil for carbon compounds produced by life, it came up empty. "Organically speaking, both samples were very clean material," said Klaus Biemann.

Joshua Lederberg, the original exobiologist, was cosmically crestfallen. "We can no longer be confident that no matter where you look you will find life," he said.

I believed with the faith of a child that more visits to Mars would soon clear up the mystery. After all, Sagan and other scientists said *Viking 1* should be the first step in a sustained search for life. But its disappointing results drained the fuel out of exobiology's rocket engines. NASA engineers built more probes to land on Mars, but they were mostly designed to work out the planet's geology and atmosphere, not to search for hints of life.

In the years after *Viking 1*, the closest thing to a search for life was an eavesdropping project. The Search for Extraterrestrial Intelligence, or SETI for short, was a NASA program that would scan the sky for radio communications from alien civilizations. JPL offered up its network of radio telescopes as our interstellar ears. Despite hostility from Congress, NASA managed to scrape together enough funds to keep developing plans for SETI through the 1980s. They even switched on their scan for a year before Congress shut down the operation.

"We shouldn't be spending precious dollars to look for little green men with misshapen heads," said Silvio Conte, a Massachusetts congressman who helped kill the project.

Just as NASA was winding down SETI, one of their scientists made a discovery in Houston that revived the world's curiosity about little green men, or at least little green microbes. The Johnson Space Center possessed a collection of meteorites, and one day in 1993 a scientist named David Mittlefehldt noticed something odd about one of them, a four-pound rock called Allan Hills 84001. It had been discovered back in 1984 by a team of geologists snowmobiling across the Allan Hills range in Antarctica. The meteorite was sitting in the middle of an ice field. It could not have eroded from the underlying ground. There were no nearby mountains from which it could have tumbled. It could only have come from the sky. After it was transported to the Johnson Space Center, scientists there identified it as a fragment of an asteroid. It sat in a nitrogen-filled cabinet for years, until Mittlefehldt got suspicious. He ran tests on the rock that revealed it didn't have the chemical signature of an asteroid. Instead, it had come from Mars.

Four billion years ago, the rock had formed on the red planet, where it had remained until an asteroid crashed into Mars, catapulting debris into space. The rock drifted for millions of years until Earth drew it into its gravity well. Thirteen thousand years ago it fell to Antarctica. It rested in the Allan Hills as the Ice Age glaciers retreated, farmers discovered agriculture, cities rose, and rockets shot into space.

When geologists finally discovered it, only eleven other Martian meteorites had ever been found. Short of sending a geologist to Mars, Allan Hills 84001 was one of the few opportunities for NASA to understand the makeup of the planet. A postdoctoral researcher named Christopher Romanek carefully inspected the rock and found blotches that suggested water had once flowed into cracks. If Mars was as warm and wet as Earth early in its history, perhaps life existed there, and perhaps life left behind microbial fossils.

Romanek joined a team of colleagues led by David McKay, and together the scientists looked for signs of life in the rock. When they bombarded bits of it with lasers, they liberated rings of carbon atoms that can form from decaying organic matter. A powerful scanning electron microscope revealed wormlike shapes that looked like cells. When McKay asked his thirteen-year-old daughter what a photograph of the worms looked like, she replied, "Bacteria." Kathie Thomas-Keprta then found crystals of magnetic minerals in the rock. On Earth, those minerals are made by bacteria, which use them as miniature compasses to help them navigate.

Had the scientists found once-living bacteria that swam in Martian oceans that dried up billions of years ago? Did Allan Hills 84001 hold proof of life? Or had they been fooled by *Eozoön* on Mars?

NASA's new definition of life didn't provide much help. McKay's team couldn't tell if the magnetic worms were self-sustained chemical systems. If they ever had been alive, they would have stopped sustaining themselves billions of years ago. As for evolution, that's something that a microbiologist can observe with a simple experiment moving slimy beads from tube to tube. But the NASA scientists had no way of tracking evolutionary change in their enigmatic structures.

The researchers chose instead to consider the ways in which both geology and biology can bring together atoms—either into lifeless minerals or into living cells. Each feature of the stony worms, taken on its own, could have formed without the presence of life. But, taken altogether, the NASA scientists decided, the evidence swung in favor of primitive cells.

When Daniel Goldin, the administrator of NASA, got wind of the study, he worried that news of it would be a disaster. Congress was fresh off of killing SETI, and a major vote on the future of NASA's funding was coming up. He called in the leaders of the project and interrogated them for hours. In the end he decided their work was sound and let it go forward. The journal *Science* accepted their paper, but it leaked before publication. As giddy speculations spread like the flu, NASA rushed to put together a press conference.

Twenty years after *Viking 1* failed to find life, even the hint of life 4 billion years ago on Mars was enough to earn some time on television news and space on front pages. President Bill Clinton even saw fit to draw attention to the discovery by issuing a statement at the White House. "If this discovery is confirmed, it will surely be one of the most stunning insights into our universe that science has ever uncovered," he said.

That prediction did not pan out. In the years after the paper was published, scientists found more evidence that lifeless chemistry could have produced the lifelike shapes. Shock waves could create magnetic minerals that looked a lot like the ones in Allan Hills 84001, for example. Twenty years after the publication of the NASA paper, the journalist Charles Choi asked a group of experts what they thought about the meteorite. Not one of them was confident that the rock contained signs of life.

Yet Allan Hills 84001 still mattered to the history of science. By making an argument that it contained fossils of once-living things, the NASA scientists focused attention on the question of life elsewhere in the universe. If Allan Hills 84001 was too ambiguous to prove life once existed on Mars, perhaps what was needed was a return mission to the planet to better understand its geology. Some researchers even pondered how they could deliver more rock back from Mars. Instead of waiting for an asteroid to blast it our way, a space probe could carefully deliver it home in pristine shape.

The debate over Allan Hills 84001 also had the good fortune of erupting just as NASA was in the midst of turning its old exobiology program into something far more ambitious. They called the new discipline astrobiology, which they defined as "the study of the living universe." To advance astrobiology, NASA supported scientists like David Deamer as they studied how life on Earth began. But they also supported research on the broad sweep of the evolution that followed, as life flooded the planet with oxygen and then produced animals, plants, and other multicellular creatures. Other astrobiologists charted the extreme forms that life takes on Earth, which might serve as analogs for alien life that could endure in extreme environments.

As for life on other planets, astrobiologists could now consider not just worlds within our own solar system but planets beyond. In 1995, Swiss researchers discovered that a sunlike star called 51 Pegasi had a tiny wobble, the gravitational tug of an orbiting planet. Since then, astronomers have discovered thousands of exoplanets of all sorts and sizes. Astrobiologists began to consider which of these exoplanets might be habitable. All life as we know it needs liquid water to survive. If a planet were too close to a hot sun, the water would boil off. Too far away, and it would freeze solid.

The more astrobiologists thought about habitability, the trickier the concept became. For one thing, a planet's habitability could change over time. In 2004, scientists at the NASA Jet Propulsion Laboratory landed a pair of rovers on Mars. As they rambled around, they came across rocks that looked as if they had formed at the bottoms of lakes and rivers long ago. If Mars wasn't habitable today, it might have been in the past.

By then the children who had watched *Viking 1*'s adventures were all grown up, paying off mortgages, and raising children of our own. We had other distractions from the heavens. Some of NASA's satellites looked down at our own planet and charted the jagged rise of its global average temperature. *Viking 1*'s children could start to see the changes for ourselves: the skating ponds that stopped freezing in winter, the king tides rolling through the Florida streets, the designer masks on sale for wildfire season.

SETI managed to emerge, phoenixlike, from congressional flames, funded by a new generation of technology tycoons. But year after year passed without any signal rising above the interstellar noise of pulsars, black holes, and leftover ripples of the Big Bang. Some researchers argued that if life is abundant on exoplanets, then SETI should have been unnecessary. Surely some intelligent aliens would have gotten in touch with us by now, either to greet or to conquer. Instead, we were surrounded by what's come to be known as the Great Silence.

By the time Laurie Barge got to the University of Southern California in 2004 to work on her PhD, she had narrowed down the scope of her curiosity to planets. "In grad school, I was obsessed with Mars," she told me.

At the time, the rovers Spirit and Opportunity were ambling around the planet. One of their stranger discoveries was blueberries—mysterious miniature blue spheres embedded in the rocky surface of Mars. Some geologists suggested the Martian blueberries had formed long ago when liquid water flowed over carbonate rocks. Barge learned how to run experiments with water and minerals to see what Martian chemistry might be capable of.

After earning her PhD, Barge got hired at JPL as a postdoctoral researcher. She worked her way up to become a co-leader of the Origins and Habitability Lab. Along the way she turned her attention and her skills with chemistry beyond Mars, to consider the possibility of life on more distant worlds: the icy moons of Saturn and Jupiter.

Galileo first spotted some of the moons of these giant planets, but it wasn't until the late 1970s that a series of JPL probes flew by them and sent back intimate portraits. Some were cratered balls of rock. Others were covered in ice. Those frozen worlds were so different from other members of the solar system that some researchers began wondering if they had conditions suitable for life.

Barge and a number of other scientists have grown especially curious about an Arizona-sized moon of Saturn called Enceladus. In 2005 the *Cassini* probe flew past the south pole of the moon and spotted a vast plume of vapor rising from giant fractures in the ice.

That surprise led the engineers at JPL to steer a new course for *Cassini*. The probe made its way back to Enceladus for a closer flyby, then back again, returning twenty-three times in total. On each visit it gulped clouds of vapor and analyzed their contents.

The plume, the scientists discovered, contains a mix of water, carbon dioxide, carbon monoxide, salt, benzene, and an assortment of other organic compounds.

That deep-space mist offered a vision of what lay below the ice. Scientists eventually concluded that its frozen shell extends down about fifteen miles, serving as a roof for a salty ocean twenty miles thick. Even though Enceladus is only 314 miles across, its ocean is much deeper than ours. The deepest point in Earth's oceans, a place called Challenger Deep, is less than seven miles.

Enceladus is 148,000 miles away from Saturn, but it takes the moon only thirty-three hours to complete an orbit around the planet. The gravitational force exerted by Saturn regularly stretches the core of Enceladus, a waterlogged ball of sand and gravel. The cycle of flexing creates enough friction to heat the core's water to a boil. It rises up to the ocean, reacting along the way with minerals and becoming a chemical-rich soup. The chill of space keeps the skin of Enceladus's ocean frozen. But Saturn's tides have broken the surface with crevices, out of which blast plumes of vapor from the warm ocean below.

Liquid water, heat, organic compounds—Enceladus has a lot of the ingredients that seem to be essential for life. In the years since *Cassini*'s visit, astrobiologists like Barge have been contemplating what organisms might lurk under the ice and how to find out if they're there. One idea is to go back to the south pole. If there are living things in the ocean of Enceladus, some of them might get shot into space in the plume. Some researchers have tinkered with nanopore sequencers to see if they could detect signs of life in the icy mist. The devices are small enough to fit aboard a space probe, and the tests astronauts have run on them aboard the International Space Station show that they can work in low gravity. On a visit to Enceladus, a probe might concentrate

DNA out of the plume and run it through a nanopore sequencer in order to read its sequence.

Life on another world might be based on DNA, or it could conceivably use another genetic molecule. If life started on Earth based on RNA, there would be no reason to rule out RNA-based life elsewhere in the universe. It was also possible that alien life used an entirely different alphabet to spell out its genetic information. Schrödinger's aperiodic crystal may take many forms that we can only barely imagine at this point. Yet, even in our ignorance, we might be able to use a nanopore sequencer to detect this alien version of life on a flight over Enceladus. If its genetic molecules are long chains encoding instructions, the sequencer may be able to slurp them through its pores and get a rough idea of their extraterrestrial scripts.

If Barge had her way, though, NASA would not give up on Enceladus after collecting some secondhand vapor. It would drop a submarine through an ice canyon into the ocean and dive to the gravelly seafloor. Barge wouldn't simply want to search for life there; she would want to explore the physical world that might have given rise to it—or might yet do so.

Such a mission may never happen. It may only get the green light after Barge has retired. Or she may get to play the part of Carl Sagan, looking in perplexity at the pictures and data that a submarine sends back from Enceladus. In the meantime Barge is content to try creating Enceladus in miniature. She builds mimics of the moon in JPL's Science Division Building.

On my visit, Barge gave me a beginner's lesson in moon building. We put on purple gloves, and then Barge handed me a vial of lime-colored crystals made of iron chloride. At her direction I poured the crystals into a clear tube of water laced with sodium silicate.

"Put a lid on it and see what happens," she said. I held the tube up at eye level. Most of the crystals fell into a heap at the bottom. After a few seconds I noticed one of them grow, stretching up like a bubble.

"Oh, you're getting a bulb—excellent!" Barge said. "That's what I wanted you to see. That's a *good* bulb—you got lucky."

The bubble stopped growing when it got to the size of a pea. Now its top began to bulge again, and a new bubble took shape. When that one stopped, another grew atop it. The heap of crystals was turning into a crooked column reaching up toward the top of the tube.

I would have felt proud of my good bulb if I knew how I had managed to make it. Instead, I asked Barge what I was seeing.

"If you could zoom in, you would see the iron crystals are dissolving," Barge said. As the iron rose from the crystals, it immediately encountered the silicate. The two substances combined and formed a porous membrane. The water trapped inside this bubble had a high pH, which caused the surrounding water to rush in through the pores. The force of the flood cracked the top of the bubble, setting the iron free to rise up farther, where it could extend the wall farther.

I was reenacting an old experiment. Alchemists who mixed together chemicals in this way described their creations as philosophical trees. It turned out that many crystals could assemble into hollow towers in water, and geochemists eventually discovered that Earth makes its own philosophical trees. Where mineral-laced water rises from the seafloor or a lake bed, it can build giant versions of the tower I had created in Barge's lab. Barge suspected that Enceladus grew its own philosophical trees in its sunless ocean.

"Basically, Enceladus at the seafloor has conditions that you could find in Earth's oceans," said Barge, "which means that the stuff you see in our oceans is possible. But in order to actually see a chimney, you'd have to go under the ice. So that's a problem."

To get a sense of what might grow on Enceladus, Barge was building her own chimneys, trying out different combinations of minerals and conditions known to produce ones on Earth. Her efforts were far more sophisticated than the little ferrous chloride bulb I made under her tutelage.

In one corner of Barge's lab, she was mimicking a 150-foot-tall tower off the coast of Iceland, known as the Strytan Hydrothermal Field. She had filled a fat syringe full of hot fluid laced with magnesium chloride that she was steadily injecting into a sealed glass bottle full of imitation seawater. A white tuft the size and shape of a baby rabbit's tail was growing inside.

Barge was planning on growing it again without oxygen, to mimic the early Earth. She had no idea what it would look like then. When she used other ingredients, she got black chimneys and others streaked in green and orange stripes. Some formed hairy plumes, and others rose like little mountains. Some turned out so strong that they stood on their own after she drained the bottles. Others fell apart like sandcastles.

"Every chimney has its own ways," Barge said.

Once Barge built her chimneys, she could run intimate experiments. She could fit them with electrodes to trace the flow of current they generate—enough in some cases to power a small LED light. In another experiment Barge and her colleagues found that amino acids, the building blocks of proteins, can form in the mineral-rich sediments that build up around chimneys as they grow.

If life exists on Enceladus, it needs a source of energy to survive. Almost a billion miles from the sun, trapped under a frozen ceiling, it cannot count on light. But Barge's research hints that it might not need it. The tidal forces acting on Enceladus ultimately produced stores of energy that life could harvest in its oceans, from hydrogen atoms cast off in chemical reactions to the electric currents generated in chimneys.

"Life can live without the sun," Barge said, "which is a big deal, because then you can have life in an ice-covered ocean."

I was struck by how Barge used the word *life* casually, without explaining what she meant by it. "Is there a definition of life that is guiding your work?" I asked.

"Not really, and I actually try not to have one," Barge said. "I've been really surprised and impressed with what organic chemistry can do when you remove life from the system. And I don't know how far it can go, honestly."

Barge's reply reminded me of the words of Stéphane Leduc, a scientist who produced dazzling philosophical trees in the early 1900s. They took on the appearance of shells, mushrooms, and flowers. Leduc believed the way his creations grew and organized themselves was not just an analogy to life. They captured some of the spirit of life. "Since we cannot distinguish the line between life and the rest of nature's phenomena, we should conclude that this line does not exist," Leduc wrote in 1910.

It's conceivable that something we'd easily recognize as life exists on Enceladus. In 2018, researchers at the University of Vienna found that a microbe that lives in the deep ocean on Earth had a metabolism that might let it survive on Saturn's moon. They re-created the ocean of Enceladus in their lab and found that the microbe could grow. But it's also conceivable that something exists on Enceladus that has no counterpart on Earth today.

Perhaps there are no microbes there. Perhaps its philosophical trees are building up a rich bounty of chemicals that gets more complex each year. They might include lipids that form oily sheets and bubbles, chains of amino acids, RNA-like threads. Enceladus might be an icy version of Darwin's warm little pond, lacking full-blown life that would otherwise feast on these chemicals. It would be wrong to call the moon's ocean a prebiotic soup, because no one can look into the future and declare that in a thousand years full-blown life will arise on Enceladus. For now, and perhaps into the future, it may hover in the borderland, where words fail.

"If we found, let's say, cells that act like cells that we observe on Earth, I'd say, 'Yeah, that's life,'" Barge told me. "If you found a lot of complex organics that look kind of biological but you don't really know how they got there, I would say, 'Maybe, but let's wait and see.' If you found physical membranes full of organics, I would be very interested to know more about them. There's a lot of stuff in between. Understanding life in the universe is about more than just finding life."

FOUR BLUE DROPLETS

Three hundred and fifty miles to the north of John Butler Burke's beaker of radiobes, and over a century forward in time, runs a strangely similar experiment. It unfolds near the river Clyde, in the Joseph Black Building at the University of Glasgow. No scientist stands over the lab bench boiling broth or purifying radium. The experiment runs itself.

It was set in motion by Lee Cronin, a chemist at the university. He and his students built a robot that could mix chemicals on its own. The robot does not walk around its lab, though. Its skeleton is a black frame anchored atop a table. Bolted to one of the frame's crosspieces is a syringe full of oil. It glides along the crosspiece until it hovers over a petri dish, into which it squirts four blue droplets. As the syringe slides away, the droplets start to move.

They glide away from each other, rushing to the sides of the dish. Slowing down, they bank away and reverse course. They head back toward their fellow droplets, but they do not collide and merge. Instead, the droplets veer away at the last minute and head off in different directions. Sometimes they twirl around each other like dancing partners and sometimes they fall into formation, traveling in circles like a school of fish in a tank.

In 1944, two psychologists named Fritz Heider and Marianne Simmel created a cartoon with triangles and rectangles they cut from a piece of cardboard. The animation began with a large triangle trapped inside a four-sided box. Heider and Simmel nudged the shapes frame by frame so that the triangle moved around inside the box until one side swung open. It then left the box and encountered a smaller triangle along with a circle.

Heider and Simmel, who both taught at Smith College, asked thirty-four of their students to watch the short movie and then write down what happened. Only one described the movie as a group of shapes moving around a frame. The rest wrote things like this:

A man has planned to meet a girl and the girl comes along with another man. The first man tells the second to go; the second tells the first, and he shakes his head. Then the two men have a fight, and the girl starts to go into the room to get out of the way and hesitates and finally goes in.

When Heider and Simmel asked another group of students to describe the personalities of the shapes, most of them picked out the same words. The big triangle was a bully, the little circle was scared, the little triangle was defiant. And when the psychologists ran the movie backwards, their students talked of different stories and different personalities.

Heider and Simmel's movie helped establish that our brains are keenly tuned to signs of life. When we see things move in complex ways, we recognize them as living things. We then quickly read their movements to figure out what their intentions are. It is so automatic that we think we are just seeing what's plainly obvious. But because it's automatic, we can't help but invest circles and triangles with life, too, even when two psychologists are nudging them around a piece of glass to create an amateur movie.

The darting droplets in Cronin's lab have the same effect on the brain. They seem at turns giddy, hesitant, gregarious, solitary.

The experience would be strange enough if Cronin were secretly steering the droplets by turning the knob on a magnetic field. But he has no control over them. His robot prepared the droplets by mixing together four simple molecules. They include octanoic acid, an ingredient in plastic. Another molecule, 1-pentanol, is made in pineapples. When the robot mixes these four chemicals together and squirts them into water, they seem to come to life, to become what John Butler Burke imagined he saw when he dropped radium into a tube of beef broth.

These blue droplets are some of the strangest residents of life's edge. Viruses may embroil scientists in debates about whether they are alive or not, but at least viruses are made of genes and proteins. Even a liposome with a few bits of RNA inside has some link to our own biology. Cronin's droplets are just crowded blobs of ordinary molecules. It's hard to find words for what the robot has created in its Glasgow lab.

When I talked with Cronin about his droplets, the word I settled on was *lifelike*. And Cronin took that as a compliment. "I would say that lifelike came before life," he said.

Biology entered the twenty-first century triumphant. While scientists had not yet found life beyond Earth, they had gotten to understand our own planet's inhabitants in luxurious detail. They knew that genes are encoded in DNA, which they could now read quickly and cheaply. They could reconstruct the genomes of Neanderthals who died 100,000 years ago. They could pick out a single cell from a drop of blood and draw up a list of every gene that was active inside it. They turned brains transparent and traced the spidery connections joining thousands of neurons in a three-dimensional network. They could find life deep underground, feeding on radioactivity. And yet all those new points of data, all those astonishing finds, did not cohere into a clear definition of life that everyone could agree on.

Living paradoxes such as viruses, mitochondria, and Amazon mollies kept getting in the way. NASA's definition of life, while memorable, didn't help NASA's own scientists as they struggled to figure out if Allan Hills 84001 contained relics of life or not. Some critics found NASA's definition not just impractical but misleading. It narrowed the possibilities that life might take.

Consider its requirement that life be capable of Darwinian evolution, for example. That's a very particular kind of change over time. It occurs when genes get precisely—but not perfectly—copied from generation to generation. Individuals with certain combinations of genes fare better at reproducing than others, and natural selection spreads those fitter versions. Over time, natural selection turns many mutations to produce new adaptations.

But can we be so sure that evolution isn't unfolding somewhere else in some other manner? Is there no chance, for example, that a different kind of biology might allow for the inheritance of acquired traits—often called Lamarckian evolution? What if heredity could flow not just between the generations but between individuals in the same generation?

This kind of dissatisfaction led to an explosion of hundreds of new definitions.

> *Life is an expected, collectively self-organized property of catalytic polymers.*
>
> *Life is a metabolic network within a boundary.*
>
> *Life is a new quality brought upon an organic chemical system by a dialectic change resulting from an increase in the quantity of complexity of the system.*

This new quality is characterized by the ability of temporal self-maintenance and self-preservation.

Life is the process of existence of open non-equilibrium complete systems that are composed of carbon-based polymers and are able to self-reproduce and evolve on the basis of template synthesis of their polymer components.

Life is a far from equilibrium self-maintaining chemical system capable of processing, transforming and accumulating information acquired from the environment.

The existence of the dynamically ordered region of water realizing a boson condensation of evanescent photons inside and outside the cell can be regarded as the definition of life.

Life is a monophyletic clade that originated with a last universal common ancestor, and includes all its descendants.

And a frank one:

Life is what the scientific establishment (probably after some healthy disagreement) will accept as life.

"It is commonly said," the scientists Frances Westall and André Brack wrote in 2018, "that there are as many definitions of life as there are people trying to define it."

As an observer of science and of scientists, I find this behavior strange. It is as if astronomers kept coming up with new ways to define stars. I once asked Radu Popa, a microbiologist who started collecting definitions of life in the early 2000s, what he thought of this state of affairs.

"This is intolerable for any science," he replied. "You can take a science in which there are two or three definitions for one thing. But a science in which the most important object has no definition? That's absolutely unacceptable. How are we going to discuss it if you believe that the definition of life has something to do with DNA, and I think it has something to do with dynamic systems? We cannot make artificial life because we cannot agree on what life is. We cannot find life on Mars because we cannot agree what life represents."

With scientists adrift in an ocean of definitions, philosophers rowed out to offer lifelines.

Some tried to soothe the debate, assuring the scientists they could learn to live with the abundance. We have no need to zero in on the One True Definition of Life, they argued, because working definitions are good enough. NASA can come up with whatever definition helps them build the best machine for searching for life on other planets and moons. Physicians can use a different one to map the blurry boundary that sets life apart from death. "Their value does not depend on consensus, but rather on their impact on research," the philosophers Leonardo Bich and Sara Green argued.

Other philosophers found this way of thinking—known as operationalism—an intellectual cop-out. Defining life was hard, yes, but that was no excuse not to try. "Operationalism may sometimes be unavoidable in practice," the philosopher Kelly Smith countered, "but it simply cannot substitute for a proper definition of life."

Smith and other foes of operationalism complain that such definitions rely on what a group of people generally agree on. But the most important research on life is at its frontier, where it will be hardest to come to an easy agreement. "Any experiment conducted without a clear idea of what it is looking for ultimately settles nothing," Smith declared.

Smith argued that the best thing to do is to keep searching for a definition of life that everyone can get behind, one that succeeds where others have failed. But Edward Trifonov, a Russian-born geneticist, wondered if a successful definition already exists but is lying hidden amidst all the past attempts.

In 2011, Trifonov reviewed 123 definitions of life. Each was different, but the same words showed up again and again in many of them. Trifonov analyzed the linguistic structure of the definitions and sorted them into categories. Beneath their variations, Trifonov found an underlying core. He concluded that all the definitions agreed on one thing: life is *self-reproduction with variations.* What NASA's scientists had done in eleven words, Trifonov now did with three.

His efforts did not settle matters. All of us—scientists included—keep a personal list of things that we consider to be alive and not alive. If someone puts forward a definition, we check our list to see where it draws that line. A number of scientists looked at Trifonov's distilled definition and did not like the line's location. "A computer virus performs self-reproduction with variations. It is not alive," declared the biochemist Uwe Meierhenrich.

Some philosophers have suggested that we need to think more carefully about how we give a word like *life* its meaning. Instead of building definitions first, we should start by thinking about the things we're trying to define. We can let them speak for themselves.

These philosophers are following in the tradition of Ludwig

Wittgenstein. In the 1940s, Wittgenstein argued that everyday conversations are rife with concepts that are very hard to define. How, for example, would you answer the question, "What are *games*?"

If you tried to answer with a list of necessary and sufficient requirements for a game, you'd fail. Some games have winners and losers, but others are open-ended. Some games use tokens, others cards, others bowling balls. In some games, players get paid to play. In other games, they pay to play, even going into debt in some cases.

For all this confusion, however, we never get tripped up talking about games. Toy stores are full of games for sale, and yet you never see children staring at them in bafflement. Games are not a mystery, Wittgenstein argued, because they share a kind of family resemblance. "If you look at them you will not see something that is common to all," he said, "but similarities, relationships, and a whole series of them at that."

A group of philosophers and scientists at Lund University in Sweden wondered if the question "What is life?" might better be answered the way Wittgenstein answered the question "What are games?" Rather than come up with a rigid list of required traits, they might be able to find family resemblances that could naturally join things together in a category we could call Life.

In 2019 they set out to find it by carrying out a survey of scientists and other scholars. They put together a list of things including people, chickens, Amazon mollies, bacteria, viruses, snowflakes, and the like. Next to each entry the Lund team provided a set of terms commonly used to talk about living things, such as order, DNA, and metabolism.

The participants in the study checked off all the terms that they believed to apply to each thing. Snowflakes have order, for example, but they don't have a metabolism. A human red blood cell has a metabolism but it contains no DNA.

The Lund researchers used a statistical technique called cluster analysis to look at the results and group the things together based on family resemblances. We humans fell into a group with chickens, mice, and frogs—in other words, animals with brains. Amazon mollies have brains, too, but the cluster analysis put them in a separate group close to our own. Because they don't reproduce by themselves, they're set a little apart from us. Further away, the scientists found a cluster made up of brainless things, such as plants and free-living bacteria. In a third group was a cluster of red blood cells and other cell-like things that can't live on their own.

Furthest away from us were things that are commonly not considered alive. One cluster included viruses and prions, which are deformed proteins that can force other proteins to take their shape. Another included snowflakes, clay crystals, and other things that don't replicate in a lifelike way.

The Lund researchers found that they could sort things pretty well into the living and the nonliving without getting tied up in an argument over the perfect definition of life. They propose that we can call something alive if it has a number of properties that are associated with being alive. It doesn't have to have all those properties, nor does it even need exactly the same set found in any other living thing. Family resemblances are enough.

One philosopher has taken a far more radical stand. Carol Cleland argues that there's no point in searching for a definition of life or even just a convenient stand-in for one. It's actually bad for science, she maintains, because it keeps us from reaching a deeper understanding about what it means to be alive. Cleland's contempt for definitions is so profound that some of her fellow philosophers have taken issue with her. Kelly Smith has called Cleland's ideas "dangerous."

Cleland had a slow evolution into a firebrand. When she enrolled in the University of California, Santa Barbara, she started off studying physics. "I was a klutz in the lab, and my experiments never turned out right," she later told an interviewer. From physics she turned to geology, and while she liked the wild places that the research took her to, she didn't like feeling isolated as a woman in the male-dominated field. She discovered philosophy in her junior year and was soon grappling with deep questions about logic. After graduating college and spending a year working as a software engineer, she went to Brown University to earn a PhD in philosophy.

In graduate school Cleland mulled space and time, cause and effect. Here's a taste of her thinking from that period:

> A dyadic relation R is *supervenient* upon a determinable non-relational attribute P if and only if
>
> 1. $\Box\ (\forall x,y)\ {\sim}\Diamond[\mathrm{R}(x,y)$ and there are no determinate attributes P_i and P_j of determinable kind P such that $P_i(x)$ and $P_j(y)]$;
>
> 2. $\Box\ (\forall x,y)\{R(x,y) \supset$ there are determinate attributes P_i and P_j of determinable kind P such that $P_i(x)$ and $P_j(y)$ and $\Box\ (\forall x,y)\ [(P_i(x)$ and $P_j(y)) \supset R(x,y)]\}$.

When Cleland finished grad school, she moved on to subjects that were easier to talk about at dinner parties. She worked at Stanford University for a time, contemplating the logic of computer programs. She then became an assistant professor at the University of Colorado, where she remained for the rest of her career.

In Boulder, Cleland turned her attention to the nature of science itself. She examined how some scientists, like physicists, could run experiments over and over again, while others, like geologists, couldn't replay millions of years of history. It was while she was reflecting about these differences that she learned about a Martian rock in Antarctica that was posing a philosophical conundrum of its own.

A lot of the arguments over Allan Hills 84001 had less to do with the rock itself than with the right way to do science. Some researchers thought the NASA team had done an admirable job of studying it, but others thought it was ridiculous to conclude from their findings that the meteorite might contain fossils. The planetary scientist Bruce Jakosky, one of Cleland's colleagues at the University of Colorado, decided to organize a public discussion where the two sides could air their views. But he realized that judging Allan Hills 84001 required more than running some experiments to measure magnetic minerals. It demanded thinking through how we make scientific judgments. He asked Cleland to join the event, to talk about Allan Hills 84001 as a philosopher.

What started as a quick prep for a talk turned into a dive into the philosophy of extraterrestrial life. Cleland concluded that the fight over Allan Hills 84001 sprang from the divide between experimental and historical sciences. The critics made the mistake of treating the meteorite study as experimental science. It was absurd to expect McKay's team to replay history. They couldn't fossilize microbes on Mars for 4 billion years and see if they matched Allan Hills 84001. They couldn't hurl a thousand asteroids at a thousand copies of Mars and see what came our way.

Cleland concluded that the NASA team had carried out good

historical science, comparing explanations for the ones that explained their evidence best. "The martian-life hypothesis is a very good candidate for being the best explanation of the structural and chemical features of the martian meteorite," she wrote in 1997 in the *Planetary Report*.

Cleland's work on the meteorite impressed Jakosky so much that he invited her in 1998 to join one of the teams at NASA's newly created Astrobiology Institute. In the years that followed, Cleland developed a philosophical argument for what the science of astrobiology should look like. She informed her ideas by spending time with scientists doing different kinds of research that fit under the umbrella of astrobiology. She traveled around the Australian outback with a paleontologist searching for clues to how giant mammals went extinct 40,000 years ago. She went to Spain to learn how geneticists sequence DNA. And she spent a lot of time at scientific meetings, roaming from talk to talk. "I felt like a kid in a candy store," she once told me.

But sometimes the scientists Cleland spent time with set off her philosophical alarms. "Everybody was working with a definition of life," she recalled. NASA's definition, only a few years old at that point, was especially popular.

As a philosopher, Cleland recognized that the scientists were making a mistake. Their error didn't have to do with determinate attributes or some other fine philosophical point understood only by a few logicians. It was a fundamental blunder that got in the way of the science itself. Cleland laid out the nature of this mistake in a paper, and in 2001 she traveled to Washington, D.C., to deliver it at a meeting of the American Association for the Advancement of Science. She stood up before an audience made up mostly of scientists, and told them it was pointless to try to find a definition of life.

"There was an explosion," Cleland recalled. "Everyone was yelling at me. It was really amazing. Everyone had their pet definitions and wanted to air them. And here I told them the whole definition project was worthless."

Fortunately, some people who heard Cleland talk thought she was onto something. She began collaborating with astrobiologists to explore the implications of her ideas. Over the course of two decades she published a series of papers, culminating in a book, *The Quest for a Universal Theory of Life*.

The trouble that scientists had with defining life had nothing to do with the particulars of life's hallmarks such as homeostasis or evolution. It had to do with the nature of definitions themselves—something that scientists rarely stopped to consider. "Definitions," Cleland wrote, "are not the proper tools for answering the scientific question 'what is life?'"

Definitions serve to organize our concepts. The definition of, say, a bachelor is straightforward: an unmarried man. If you're a man and you're unmarried, you are—by definition—a bachelor. Being a man is not enough to make you a bachelor, nor is being unmarried. As for what it means to be a man, well, that can get complicated. And marriage has its own complexity. But we can define "bachelor" without getting bogged down in those messy matters. The word simply links these concepts in a precise way. And because definitions have such a narrow job to do, we can't revise them through scientific investigation. There is simply no way that we could ever discover that we were wrong about the definition of a bachelor as being an unmarried man.

Life is different. It is not the sort of thing that can be defined simply by linking together concepts. As a result, it's futile to search for a laundry list of features that will turn out to be the real definition of life. "We don't want to know what the word *life* means to us," Cleland said. "We want to know what life *is*." And if we want to satisfy our desire, Cleland argues, we need to give up our search for a definition.

Before the modern age of chemistry, alchemists tried to define water in the same way many biologists define life: by putting together a list of its qualities. Water is a liquid, it's clear, it's a solvent that breaks up other substances, and so on. Far from clearing up the mystery of water, however, this definition only got the alchemists into more trouble when they discovered that not all water is alike. Some kinds of water dissolved different substances and not others. So the alchemists gave different names to those waters. But then they got into even more trouble with their definitions when they watched water freeze or boil. Ice and vapor do not share the properties of liquid water. The alchemists were forced to declare that they were entirely different substances.

The quandary was so deep that it left even Leonardo da Vinci confused:

> And so [water] is sometimes sharp and sometimes strong, sometimes acid and sometimes bitter, sometimes sweet and sometimes thick or thin, sometimes it is seen bringing hurt or pestilence, sometimes health-giving, sometimes poisonous. So one would say that it suffers change into as many natures as are the different places through which it passes. And as the mirror changes with the colour of its object so it changes with the nature of the place through which it passes: health-giving, noisome, laxative, astringent, sulphurous, salt, incarnadined, mournful, raging, angry, red, yellow, green, black, blue, greasy, fat, thin.

Crafting a new definition of water wouldn't have freed Leonardo from his ignorance. The difficulty lay elsewhere: in just how little he and everyone else in the Renaissance knew about chemistry.

It took three centuries for a mature theory of chemistry to emerge, one that explained that the universe is made up of atoms belonging to many elements, which can bond together to form different molecules. Water, once thought to be an element, turned out to be made of molecules that were a combination of two elements: a pair of hydrogen atoms and an oxygen atom. These molecules made up the liquid water in a lake as well as the water in a block of ice or in a cloud of vapor. Chemists also determined that strong water and noble water were not water at all, since they were made up of different molecules.

But even H_2O is not the definition of water. A single molecule cannot do the things that water does. When water freezes, for example, it expands as many molecules of H_2O spontaneously lock into a crystal lattice. "Talking about water being H_2O tells you nothing about that," said Cleland. Instead, knowing about H_2O opens the way to learning more about the nature of water.

When it comes to life, Cleland argues, we are still alchemists. We use our intuitions to decide which things are alive or not and make arbitrary lists of the features they share. We paper over our ignorance with definitions that never manage to capture the thing we're trying to understand. The best thing that scientists can do right now, Cleland argues, is work toward a theory that explains life.

I've met many scientists who agree with Cleland on this point. They do not have a theory of life yet. They're confident that someday a theory will emerge, but for now they can only guess what it will be. It's as if they're reading the shadows that the theory is casting back to us from the future. I once asked a biophysicist to describe what the theory would sound like. "This is what life *has* to be," he replied.

Theories don't pop into existence. They only emerge after scientists have carried out lots of tedious measurements of the world. The architects of modern chemistry ran countless experiments to determine the ratios that made up compounds such as water. They discovered that the ratios were simple, made up of integers. Water was made up of two parts hydrogen to one part oxygen. Methane was four parts hydrogen to one part carbon. Out of this painstaking accounting emerged the profound realization that these compounds were molecules made out of atoms.

Some scientists today believe that a theory of life can emerge only from exacting measurements of living things. They are inventing tools to precisely measure the timing by which genes turn on and off, the rate at which cells grow, the interconnected links by which living things sense the world and make decisions about what to do next. It may take decades before these precise measurements reveal patterns that let scientists recognize a full-blown theory.

Other scientists are not so patient. They've created theories that explain life based on what scientists have already discovered. Even a simple precursor of a full-blown theory may be useful, they argue, if only to give scientists an idea of what they need to measure to build a better theory.

The first theories of life took shape once molecular biologists worked out some of the basic rules of DNA and protein in the mid-1900s. Only a few scientists dared to build theories at first, and they worked mostly in obscurity. It was an obscurity partly of their own making. They invented personal languages to think through their ideas, which they didn't put much effort into helping other people understand. The story goes that two of these theorists, Robert Rosen and Francisco Varela, met each other once at a scientific meeting. They couldn't think of a single thing to say to each other.

For all their mutual incomprehension, life's theorists worked in much the same way. They developed compressed descriptions of life that could account for the patterns seen in living things. To do so required looking beyond the marvels and puzzles of pythons and slime molds, to see the essential conditions for something to stay alive. It was as if they were physicists encountering airplanes for the first time. If they wanted to figure out how planes fly, it would have been a waste of time to study a modern airliner. They'd get lost in video screens, call buttons, and snack carts. To discover the things that matter to flight itself, they'd be better off going to Kitty Hawk and studying the *Wright Flyer*, with its simple wings of spruce and ash.

In the 1960s a medical student named Stuart Kauffman joined this tiny society. At the time, biologists were discovering some of the deep connections between genes and proteins that make life possible. They were finding that certain genes become active only if a certain protein lands on the DNA nearby. They found some of the links in the long chains of reactions that make metabolism possible. Kauffman wondered if some basic principles lurk under the dizzying details of particular proteins in particular species.

Kauffman developed a kind of algebra for the cell, which he used to create hypothetical genes and proteins on a computer. In one experiment he tried to build a simple metabolism. For food he created two molecules: Call them A and B. An A and a B had a certain probability of joining together to make a bigger molecule, AB. And AB in turn had a certain probability of combining to make even bigger molecules. Add an extra B to make ABB; combine two ABs to make ABAB. While Kauffman's metabolism could build up bigger molecules, he also programmed it so that it could break some of the bigger ones back down into fragments.

Using different rules for building and breaking molecules, Kauffman tested out a number of networks. Most of them failed to do much of anything. They only managed to use the As and Bs he fed them to make small molecules but never any big ones. But every now and then he found a network that seemed to come to life. In these networks Kauffman discovered that a few of the possible molecules became abundant. Once those few molecules grew common, they stayed common as long as Kauffman kept feeding the network.

Kauffman discovered that the successful molecules had joined together into loops of chemical reactions. One molecule would spur the growth of a second, which spurred the growth of a third, and so on until the last molecule in the loops, which helped the first. As each molecule became more abundant, it could help build its partners in a self-sustaining cycle.

Kauffman named these networks autocatalytic sets. The name refers to catalysts, which are any sort of substance that speeds up a chemical reaction between two other substances. Enzymes are just one kind of catalyst, as are certain kinds of metals. In cars, for example, platinum acts as a catalyst to break down exhaust in catalytic converters. Oil is the product of catalysts deep under the ocean floor. Autocatalytic sets were different from ordinary catalysts, Kauffman argued, because they catalyze each other.

Even though he discovered them on a computer, Kauffman

became convinced that autocatalytic sets captured something essential about life. Living things sustain themselves with networks of real molecules, he suggested. A theory of life based on autocatalytic sets would have no need for a mystical vital force giving life to lifeless matter. When Kauffman built random networks, the autocatalytic sets spontaneously took shape within them.

By the 1980s a number of other scientists picked up Kauffman's ideas about autocatalytic sets. His theory proved a useful guide to thinking about life. But the only place where scientists could watch autocatalytic sets in action was on a computer, where networks could feed on digital grub. Eventually, though, chemists succeeded in crafting autocatalytic sets out of actual molecules rather than ones and zeros. One of the most complex autocatalytic sets was built by Reza Ghadiri, a chemist at Scripps Research Institute. He and his colleagues used small chains of amino acids called peptides. They designed a set of peptides that could line up peptide fragments and bond them together. After mixing together dozens of different kinds of peptides and fragments, the scientists stepped back and let them mingle. An autocatalytic set spontaneously emerged, made of nine peptides that could build each other out of the fragments, multiplying into millions of new copies.

Autocatalytic sets are not just mathematical dreams, it turns out. But that doesn't mean they're common in nature. A mixture of chemicals is far more likely to just reach an equilibrium and do nothing more. Why autocatalytic sets only arise rarely remains an open question. They may require a supply of molecules that are in just the right proportion. Otherwise they can't build up enough new molecules to sustain the right reactions. It's also possible that autocatalytic sets are rare because they are prone to collapsing. Only if they have a resilient structure—perhaps loops within its loops—can they withstand hard times when their ingredients run low.

Scientists will have to settle questions such as these before autocatalytic sets can become part of a mature theory of life. Such a theory might explain how life sustains itself and perhaps even how it emerged in the first place. In 2019, Stuart Kauffman and two colleagues considered David Deamer's scenario in which life started as RNA-based protocells in drying ponds. They made some rough estimates of the variety of RNA molecules that could have formed in such a pond. Kauffman and his colleagues concluded that a single pool could very well have produced an autocatalytic set of RNA molecules. Once this kind of self-sustaining chemistry got started, it could then evolve into living things. Before there was life, in other words, there may have been autocatalytic sets.

Living things are special, but they're not the only special things in the universe. In 1911 the Dutch physicist Heike Kamerlingh Onnes discovered that a mercury wire chilled down to close to absolute zero becomes very special indeed. At ordinary temperatures a current loses some energy as it travels down a metal wire—a property called resistance. When Onnes cooled his mercury wire in a bath of liquid helium, the resistance gradually dropped until it reached –452°F. Then, suddenly, its resistance fell to zero. If he fashioned a loop of metal, a current could travel around it indefinitely without any loss at all.

"Mercury has passed into a new state," Onnes declared, "which on account of its extraordinary electrical properties may be called the superconductive state."

Onnes later found that other metals, like tin and lead, could also enter this new state when they got close to absolute zero. Certain mixtures of metals could become superconductive at warmer temperatures. Physicists searched for superconductivity in all its forms, eager to find materials to build fundamentally new kinds of technology. But their research remained little more than a game of trial and error for decades. Ordinary physics seemed unable to explain it, and there was no rhyme or reason to why some substances were superconductive and not others.

Albert Einstein tried to explain superconductivity with an elegant theory, which turned out to be wrong. So did Niels Bohr, Richard Feynman, and other leading figures in twentieth-century physics. Finally, in the 1950s, John Bardeen, Leon N. Cooper, and Robert Schrieffer came up with a theory that made sense of the senseless. Resistance is the result of electrons hopping around in a disorderly way, sending off the energy of a current in all directions. Bardeen, Cooper, and Schrieffer argued that some of the electrons in superconducting material form pairs that travel along the same path. Their order counteracts the chaos in a conducting metal, wiping away all resistance to a current. The new theory of superconductivity explained why some metals entered the strange state and not others, and it helped bring this special state of matter closer to our ordinary lives—in magnets that support high-speed trains and in microprocessors that may become the brains of a new generation of computers.

A theory of life may end up looking a lot like the theory of superconductivity. It may explain life as a particular configuration of matter that gets a special quality from the physics of the universe. Lee Cronin has been working with Kate Adamala and Sara Walker, a physicist at Arizona State University, on an explanation of life as a special way of putting things together. They call it assembly theory.

You can think of the history of the universe as 13.7 billion years of things being put together. After the Big Bang, subatomic particles formed hydrogen atoms; helium atoms came about as hydrogen atoms joined together. Stars were assembled from hydrogen and helium, and in their stellar forges, new elements formed. Atoms assembled into molecules; molecules became grains. Planets and moons were built from them. On Earth snowflakes formed in the sky. Underground, minerals took shape.

Once life emerged, it made things of its own. Organisms began making sugars, proteins, and cells. They grew tusks and flowers. Animals constructed beehives, beaver lodges, double-hulled canoes, and space probes. Cronin, Adamala, and Walker worked with colleagues on an objective way to compare how things get assembled, whether life is involved or not.

The assembly of things happens in steps. A simple molecule may need just a single step to form from atoms. But it takes more steps to add extra atoms or to join two molecules together. Cronin and his colleagues figured out a way to estimate the number of steps it takes to make a molecule: by smashing it. Think of molecules as Lego constructions that you take apart at random. If someone gives you a hundred constructions that are made of just two Lego blocks snapped together, you will be able to break them only into the same two blocks over and over again. But now imagine someone gives you a Lego Hogwarts Castle, complete with turrets, buttresses, and archways. It can be split into many different fragments. Cronin and his colleagues found that the number of fragments a molecule can be broken down to is a good guide to the number of steps it takes to build it in the first place.

Cronin and his colleagues carried out a survey by smashing,

breaking apart over a hundred different materials. They smashed quartz and limestone. They broke apart Taxol, a molecule made by yew trees that turns out to be a powerful drug for cancer. They followed Stanley Miller's recipe and made a prebiotic soup and then broke its molecules back down. They broke down beer and granite.

All the materials that were not made by living things needed under fifteen steps to assemble, the researchers determined. Even when they experimented with a tiny sample of the Murchison meteorite—packed with lipids, amino acids, and other building blocks of life—they never found a molecule that needed fifteen steps. "Although there's a billion molecules in it, it's all boring," Cronin told me.

Living things, by contrast, were not boring. Although they assembled some simple molecules, they also made exquisitely complex ones, some of which needed far more than fifteen steps.

It's possible assembly theory has revealed a line cutting through life's borderland. Ordinary chemistry may not be able to assemble a material that needs fifteen steps or more. Any one reaction may take place, given enough time. But the odds may be vanishingly small that a series of certain reactions happens in the right order—and happens in that order again and again. Life, on the other hand, is a state of matter that can spontaneously make things with a lot of assembly steps.

Adamala, Cronin, and Walker have proposed that what allows life to do this is the special way in which information flows through it. In life, information is able to control matter. Genes and other molecular structures can store information, copy it into their offspring, and then channel that information through networks of proteins in order to do precise tasks—such as making things in many steps of assembly.

Assembly theory might offer a way to look for life on other planets. It might be possible to detect life on planets orbiting other stars without even visiting them. Astronomers could use telescopes to scan the atmospheres of exoplanets for molecules. If they detect a molecule with a high assembly number in abundance, they can be confident that it didn't come into being through random chemistry. Only information could guide its production.

But Cronin doesn't have to wait for a billion-dollar probe to get to the other side of the solar system or for a new space telescope to go up into geosynchronous orbit. "I can now go look for lifeforms in my lab," he said. "Forget whether life is like a flame, forget whether it has a metabolism. Does the object have enough features to say that it can't have formed randomly? And do you find it in abundance? If yes, then it's alive. If no, you can't tell whether it was alive or not."

Cronin chose droplets as his raw material for making life. Compared to David Deamer's liposomes, they're even simpler—just blobs of oil. They crowd together thanks to the difficulty they have bonding with water molecules. Mixing other chemicals into the oil can make droplets do intriguing things. Alcohol, for example, is attracted to both oil molecules and water molecules. When Cronin mixes it into a droplet, it will leak out slowly. As each alcohol molecule leaves, it gives the droplet a little push, moving it in the opposite direction. If enough alcohol flows out, the droplet will look as if it's swimming. Different mixtures of chemicals will make droplets behave differently. Even a slight twist to a recipe can lead to unexpected behaviors.

To explore the universe of possibilities, Cronin realized that he and his team couldn't just carry out experiments by hand. They had to build a robot to run them. Cronin christened it the DropFactory. The DropFactory began running thousands of experiments in a row. To make droplets that could race, it started by mixing four oils together and pouring droplets into petri dishes. It then moved the dishes under a video camera to film them, washed out the dishes, and then mixed the oils into a new combination. Some recipes made droplets that didn't budge, and others made them move faster. The DropFactory used these results to make a model of the chemistry in the droplets, which it updated with each new experiment. At the end of this robotic evolution, the droplets were racing like puppies set loose in a dog run.

The DropFactory can learn to make droplets that do other things. It has come up with a recipe that made the droplets wobble as if Glasgow were the epicenter of an earthquake. In another trial, the droplets gained the ability to split in two, producing a nest of little offspring. Cronin's team programmed the robot with curiosity, so that it would notice odd new behaviors on its own and make them stronger. The robot discovered a recipe for droplets that dawdled at room temperature and then took off in a sprint when the air got only a few degrees warmer.

These lifelike droplets, these skittering blobs of active matter, are not life. But they could be a dry run for making it. Into these droplets Cronin plans to introduce more chemicals—sugars, pyrite, silicates—that another robot in his lab is busy creating through its own rounds of chemical reactions.

Cronin hopes that his robots will ultimately create droplets that are capable of behaving in complex ways and can also carry out complex chemistry to make new compounds. It's possible that this prebiotic evolution will favor chemicals that can store information and pass it along when a single droplet splits in two. Cronin hopes the droplets will mimic Kauffman's autocatalytic sets, cooperating to carry out complex chemistry that would be too much for one droplet to handle on its own.

Assembly theory could potentially let Cronin and his colleagues make sense of these changes in the droplets. If they start assembling chemicals that couldn't be produced without the information-guided processes, then the scientists can declare the droplets alive. Cronin wouldn't be surprised if the chemistry that brings them to life has nothing to do with life as we know it, based on DNA or even RNA. "It's like saying gravity only works on *this* kind of rock," he said.

Cronin knows that a lot of scientists are skeptical that a robot can make life from simple chemicals. Surely, it must have taken a long time for an entire planet to get the job done. But Cronin thinks that has to be wrong. Primordial molecules were too fragile to sit around for very long. If life formed, it had to form fast.

"Just the back-of-an-envelope calculation tells me ten thousand hours," Cronin said. "I'm pretty sure we will crack the origin-of-life problem in the next few years. But then everyone will go, 'Oh, that was easy.'"

Cronin's confidence was at once so sincere and so strange that I began planning a trip to Glasgow in a decade to visit his lab—either to admire his flock of living droplets, or to observe what happens when radiobes get the better of a scientist once more.

"I'm either mad," Cronin declared, "or completely right."

NOTES

Introduction: The Borderland

ix Cavendish Laboratory: Cavendish Library 1910; Thomson 1906.

ix "the most primitive form of life": Quoted in the *Guardian* 1905, p. 6.

ix John Butler Burke: Biographical details are drawn from "Mr. J. B. Butler Burke" 1946; Burke 1906; Burke 1931a; Burke 1931b; Campos 2006; Campos 2007; Campos 2015; "A Filipino Scientist" 1906.

x "a man who is gifted with": Quoted in Burke n.d.

x "an old lady of very large means": Ibid.

xi "Are we about to realize": Quoted in Badash 1978, p. 146.

xi "It changes its substance": Quoted in Burke 1903, p. 130.

xii "They are entitled to be classed": Quoted in Burke 1905b, p. 398.

xiii "The constituents of protoplasm are": Quoted in Burke 1906, p. 51.

xiii They belted out "The Radium Atom": Satterly 1939.

xiv published his first report on radiobes: Burke 1905a.

xiv "Life may have originated": Quoted in "The Origin of Life" 1905, p. 3.

xiv "Has Radium Revealed the Secret of Life?": Hale 1905.

xiv "John Butler Burke has suddenly become": Quoted in "The Cambridge Radiobes" 1905, p. 11.

xiv "one of the most brilliant": Quoted in "City Chatter" 1905, p. 3.

xv "Mr. Burke attained suddenly to a notoriety": Quoted in Campbell 1906, p. 89.

xv "provoked more discussion": Quoted in "A Clue to the Beginning of Life on the Earth" 1905, p. 6813.

xv *The Origin of Life*: Burke 1906.

xv "perception in the universal mind": Quoted ibid., p. 345.

xv "Biology is decidedly not his forte": Quoted in Campos 2006, p. 84.

xvi "mere drawing": Quoted in Douglas Rudge 1906, p. 380.

xvi "Mr. Rudge has carried out the experiments": Quoted in Campbell 1906, p. 98.

xvii "The Radiobe": Satterly 1939.

xvii "the dreaded event of bankruptcy": Quoted in Burke n.d.

xviii "Burke had gone right off the deep end": Quoted in Campos 2015, p. 96.

xviii "Life is what IS": Ibid.

xix life's edge: Cornish-Bowden and Cárdenas 2020.

Part One: The Quickening

The Way the Spirit Comes to the Bones

5 elk kelp: Herbst and Johnstone 1937.

7 Alysson Muotri: The research I discuss in this chapter can be found in Marchetto et al. 2010; Cugola et al. 2016; Mesci et al. 2018; Setia and Muotri 2019; and Trujillo et al. 2019.

8 cerebral cortex: Stiles and Jernigan 2010.

8 brain organoids: Lancaster et al. 2013.

16 "We know what it feels like": Quoted in Haldane 1947, p. 58.

16 Cotard's syndrome: Berrios and Luque 1995; Berrios and Luque 1999; Dieguez 2018; Cipriani et al. 2019.

16 "affirms she has no brain": Quoted in Debruyne et al. 2009, p. 197.

17 She refused to bathe: Debruyne et al. 2009.

17 turned him into a zombie: Huber and Agorastos 2012.

17 In 2015, Indian doctors: Chatterjee and Mitra 2015.

17 unconscious shortcuts: Rosa-Salva, Mayer, and Vallortigara 2015.

18 biological or physical motion: Caramazza and Shelton 1998.

18 In one series of experiments: Fox and McDaniel 1982.

18 Brain damage: Moss, Tyler, and Jennings 1997.

18 look longer at geometrical shapes: Bains 2014; Di Giorgio et al. 2017.

19 "If we carve the human mind": Quoted in Nairne, VanArsdall, and Cogdill 2017, p. 22.

19 Experiments on animals: Anderson 2018; Gonçalves and Carvalho 2019.

19 Giorgio Vallortigara and Lucia Regolin: Vallortigara and Regolin 2006.

19 visual shortcuts: Connolly et al. 2016.

21 "ensoulment": Neaves 2017.

21 The Beng: Gottlieb 2004.

21 "Although he impedes the ensoulment": Quoted in Noonan Jr. 1967, p. 104.

22 "Life is the immediate gift of God": Quoted in Blackstone 1765, p. 88.

23 "It is a sin against God": Quoted in Peabody 1882, p. 4.

25 when an embryo becomes a person: Manninen 2012.

25 "Life begins at conception": Quoted in Berrien 2017.

25 "There is no single, simple answer": Quoted in Lederberg 1967, p. A13.

26 mixed their molecules together: Rochlin et al. 2010; Aguilar et al. 2013.

27 "A distinct, living human individual": Quoted in Lee and George 2001.

27 makes it impossible to pin: Peters Jr. 2006.

28 make a fresh batch of proteins: Vastenhouw, Cao, and Lipshitz 2019; Navarro-Costa and Martinho 2020.

29 when it became two people: Devolder and Harris 2007; Rankin 2013.

29 chimeras: Maienschein 2014.

29 development often ends in failure: Giakoumelou et al. 2016; El Hachem et al. 2017; Vázquez-Diez and FitzHarris 2018.

30 One study published in 2016: Jarvis 2016a; Jarvis 2016b.

30 this crisis: Simkulet 2017; Nobis and Grob 2019.

31 these lives couldn't be saved anyway: Blackshaw and Rodger 2019.

31 hormone injections: Haas, Hathaway, and Ramsey 2019.

31 "In the situations of rape and incest": Quoted in WFSA Staff 2019.

32 a reprogrammed cell: Ball 2019.

33 the integration of information: Koch 2019a.

34 decide how we'll care: Hostiuc et al. 2019.

34 "What would it mean for": Quoted in Koch 2019b.

Death Is Resisted

35 James Forbes: Dyson 1978.

35 "the TRULY SPLENDID work before us": Quoted in "Oriental Memoirs" 1814, p. 577.

36 "When convinced that": Forbes 1813, p. 333.

36 the lamentably moaning monkey: Wakefield 1816; Gulliver 1873.

37 "So intense is the grief": Quoted in Darwin 1871, p. 48.

37 Jane Goodall: Van Lawick-Goodall 1968; Van Lawick-Goodall 1971.

39 even fish: Gonçalves and Biro 2018; Gonçalves and Carvalho 2019.

39 *cadaverine* and *putrescine*: Samartzidou et al. 2003; Hussain et al. 2013; Crippen, Benbow, and Pechal 2015.

39 primate thanatology: Gonçalves and Carvalho 2019.

40 a modern sense of death: Hovers and Belfer-Cohen 2013; Pettitt 2018.

41 Buddhists: Bond 1980; Simpson 2018.

42 "Medical men have rarely discussed": Quoted in Ackerknecht 1968, p. 19.

42 Xavier Bichat: Bichat 1815; Haigh 1984; Sutton 1984.

42 "consists in the sum of the functions": Quoted in Bichat 1815, p. 21.

43 a blurry boundary: I draw my account of Van Leeuwenhoek and Needham from Keilin 1959 and Clegg 2001.

45 "We find an Instance here": Quoted in Baker 1764, p. 254.

46 grew into healthy new plants: Yashina et al. 2012.

47 "a third state between life and death": Quoted in Cannone et al. 2017, p. 1.

48 "the hard backup of this planet": Quoted in Oberhaus 2019.

48 without breath or heartbeat: Bondeson 2001.

49 "It was a grisly place": Quoted ibid., p. 109.

50 iron lungs: Slutsky 2015.

51 mixed blessing: Vitturi and Sanvito 2019.

51 "a new, previously undescribed, state": Quoted in Goulon, Babinet, and Simon 1983, p. 765.

51 *coma dépassé*: Mollaret and Goulon 1959.

51 Other doctors agreed: Wijdicks 2003.

51 "The developments in resuscitative and supportive therapy": Quoted ibid., p. 971.

52 "Patients are being brought in": Ibid., p. 972.

53 transplanted kidney: Machado 2005.

53 the death of the brain: Beecher 1968.

54 The committee's report: Bernat 2019.

54 "Harvard Panel Asks Definition of Death Be Based on Brain": Reinhold 1968.

55 "The inescapable logic of the concept": Quoted in Sweet 1978, p. 410.

55 "whole-brain standard": Quoted in President's Commission for the Study of Ethical Problems in Medicine and Biomedical and Behavioral Research 1981.

56 “Plaintiffs are Christians with”: Quoted in Aviv 2018.

56 “She is going to start to decompose”: Quoted in Szabo 2014.

58 “I am convinced that”: Quoted in Shewmon 2018, p. S76.

58 “Perhaps McMath actually improved”: Quoted in Truog 2018, p. S73.

59 1,800 patients: Nair-Collins, Northrup, and Olcese 2016.

59 Michael Nair-Collins: Nair-Collins 2018.

61 published his first defense of brain death: Bernat, Culver, and Gert 1981.

61 “may represent a false-positive determination”: Quoted in Bernat and Ave 2019, p. 636.

61 “Death is a biological and irreversible event”: Quoted in Huang and Bernat 2019, p. 722.

61 Jahi began to falter: Dolan 2018.

61 “I told her”: Quoted in Ruggiero 2018.

Part Two: The Hallmarks

Dinner

65 For snake metabolism, see Diamond 1994; Secor, Stein, and Diamond 1994; Secor and Diamond 1995; Secor and Diamond 1998; Andrew et al. 2015; Larsen 2016; Andrew et al. 2017; Engber 2017; Perry et al. 2019.

66 they experience a red-out: Boback et al. 2015; Penning, Dartez, and Moon 2015.

Decisive Matter

80 For slime molds and intelligence, see Brewer et al. 1964; Ohl and Stockem 1995; Dussutour et al. 2010; Reid et al. 2012; Reid et al. 2015; Adamatzky 2016; Reid et al. 2016; Oettmeier, Brix, and Döbereiner 2017; Boussard et al. 2019; Gao et al. 2019; Ray et al. 2019.

Preserving Constant the Conditions of Life

93 For bats and homeostasis, see Webb and Nicoll 1954; Adolph 1961; McNab 1969; Cryan et al. 2010; Pfeiffer and Mayer 2013; Hedenström and Johansson 2015; Johnson et al. 2016; Boyles et al. 2017; Voigt et al. 2017; Willis 2017; Bandouchova et al. 2018; Gignoux-Wolfsohn et al. 2018; Moore et al. 2018; Boerma et al. 2019; Boyles et al. 2019; Haase et al. 2019; Rummel, Swartz, and Marsh 2019; Auteri and Knowles 2020; Lilley et al. 2020.

Copy/Paste

102 For maple trees, see Taylor 1920; Peattie 1950; De Jong 1976; Green 1980; Stephenson 1981; Sullivan 1983; Hughes and Fahey 1988; Burns and Honkala 1990; Houle and Payette 1991; Peck and Lersten 1991a; Peck

and Lersten 1991b; Graber and Leak 1992; Greene and Johnson 1992; Greene and Johnson 1993; Abrams 1998.

105 It's far less true for slime molds: Clark and Haskin 2010.

Darwin's Lung

112 *Pseudomonas aeruginosa*: Moradali, Ghods, and Rehm 2017.

114 "We see nothing of these slow changes": Quoted in Zimmer 2011, p. 42.

116 Scores of graduate students: Poltak and Cooper 2011; Flynn et al. 2016; Gloag et al. 2018; Gloag et al. 2019.

116 run a weeklong experiment: Cooper et al. 2019.

116 *Pseudomonas fluorescens*: Ferguson, Bertels, and Rainey 2013.

122 "bacteria of blue pus": Quoted in Villavicencio 1998, p. 213.

Part Three: A Series of Dark Questions

This Astonishing Multiplication

127 For Trembley, see Baker 1949; Vartanian 1950; Baker 1952; Beck 1960; Lenhoff and Lenhoff 1986; Dawson 1987; Lenhoff and Lenhoff 1988; Dawson 1991; Ratcliff 2004; Baker 2008; Stott 2012; Gibson 2015; Steigerwald 2019.

128 "Nature must be explained":Quoted in Lenhoff and Lenhoff 1988, p. 111.

129 René Descartes: Dawson 1987; Slowik 2017.

129 Friedrich Hoffmann: Quoted from Hoffman 1971, p. 6.

129 anti-Cartesians: Roe 1981, p. 107.

130 "Above all else, consequently": Quoted in Zammito 2018, p. 24; Beck 1960.

130 a soul: Zammito 2018, p. 25.

131 "The more I followed": Quoted in Lenhoff and Lenhoff 1986, p. 6.

132 "roused sharply in my mind": Ibid.

133 "I am entirely taken up": Quoted in Ratcliff 2004, p. 566.

134 the stuff of life: Baker 1949.

135 "appeared so extraordinary": Quoted in Baker 1743.

135 "A miserable insect has just shown": Quoted in Dawson 1987, p. 185.

Irritations

136 For Haller, see Reed 1915; Haller and Temkin 1936; Maehle 1999; Lynn 2001; Steinke 2005; Frixione 2006; Hintzsche 2008; Rößler 2013; Cunningham 2016; McInnis 2016; Gambarotto 2018; Zammito 2018; Steigerwald 2019.

136 "Called to Göttingen": Quoted in Cunningham 2016, p. 95.

137 "To go through everything": Quoted in Cunningham 2016, p. 93.

138 "One needs a stock": Quoted in Steinke 2005, p. 53.

138 "a species of cruelty": Quoted in Haller and Temkin 1936, p. 2.

140 "the most irritable organ of all": Ibid., p. 53.

140 "I saw in all corners": Quoted in Steinke 2005, p. 136.

140 "The enemies of Mr. Haller": Quoted in Zammito 2018, p. 75.

141 "What to answer to 1,200 experiments?": Quoted in Steigerwald 2019, p. 66.

141 "the pain became unbearable": Quoted in Rößler 2013, p. 468.

142 "candor and love of truth": Quoted in Maehle 1999, p. 159.

142 "terrible weakness of mind": Ibid., p. 183.

142 "with the hush of the winds": Quoted in Hintzsche and Wolf 1962.

143 "*It no longer beats—I die*": Quoted in Reed 1915, p. 56.

The Sect

144 "The vanity of attempting": Haller and Temkin, p. 2.

144 Buffon: Roger 1997.

145 "I confess that I can only": Quoted in Baker 1952, p. 182.

145 "Irritability is becoming a sect": Quoted in Zammito 2018, p. 89.

145 "lies a special, innate effective drive": Quoted in Steigerwald 2019, p. 86.

147 "We are full of horror": Quoted in King-Hele 1998, p. 175.

149 define water: Cleland 2019a.

150 "In living Nature": Quoted in Hunter 2000, p. 56.

150 "I can no longer": Ibid.

151 "The natural dividing wall": Quoted in Ramberg 2000, p. 176.

This Mud Was Actually Alive

152 Lord George Granville Campbell: For the *Challenger* voyage, see Campbell 1877; Macdougall 2019.

152 "had dropped down on the ocean": Quoted in Campbell 1877, p. 39.

153 "The name came out": Quoted in Moseley 1892, p. 585.

154 Thomas Huxley: Geison 1969; McGraw 1974; Rehbock 1975; Rupke 1976; Rice 1983; Welch 1995; Desmond 1999.

156 "kind of soft, mealy substance": Quoted in Huxley 1868, p. 205; McGraw 1974.

157 "If the hypothesis of evolution": Quoted in Huxley 1891, p. 596.

158 primal slime: Ibid.

158 the same living jelly: Geison 1969.

158 "The cell was redefined": Liu 2017.

158 "I dare not venture": Ibid., p. 912.

159 "a little particle of": Quoted in Carpenter 1864, p. 299; Burkhardt et al. 1999.

159 *Eozoön*: O'Brien 1970; Adelman 2007.

159 "should not be astonished": Quoted in King and Rowney 1869, p. 118.

160 "deep-sea 'Urschleim'": Quoted in Huxley 1868, p. 210.

160 "I hope that you will not": Quoted in Rehbock 1975, p. 518.

160 "living paste": *Athenaeum* 1868.

161 the physical basis of life: Geison 1969; Huxley 1869.

161 "The audience seemed almost": Quoted in Hunter 2000, p. 69.

162 "This mud was actually alive": Quoted in Thomson 1869, p. 121.

162 "It probably forms one continuous scum": Quoted in Rupke 1976, p. 56.

162 "fanciful and improbable": Quoted in Beale 1870, p. 23.

163 a zoology textbook: Packard 1876.

163 "huge masses of naked, living protoplasm": Quoted in Rehbock 1975, p. 522.

164 "If the jelly-like organism": Quoted in Buchanan 1876, p. 605.

165 "In placing it amongst living things": Quoted in Murray 1876, p. 531.

165 "deny that such a thing exists": Quoted in Rehbock 1975, p. 529.

165 "I am mainly responsible": Quoted in Huxley 1875, p. 316.

166 "The more the real parent": Quoted in Rehbock 1975, p. 531.

166 "a case in which": Quoted in McGraw 1974, p. 169.

166 "*Bathybius* was a highly functional concept": Quoted in Rupke 1976, p. 533.

167 "Whatever bit of life he touched": Quoted in "Obituary Notices of Fellows Deceased" 1895.

A Play of Water

168 making beer: Liu et al. 2018.

170 An ordinary enzyme: Barnett and Lichtenthaler 2001.

170 "will enjoy none too long a life": Quoted in Kohler 1972, p. 336.

170 "The differences between": Quoted in Buchner 1907.

171 "Which of them are alive?": Quoted in Wilson 1923, p. 6.

171 more than one level: Nicholson and Gawne 2015.

172 a religious phenomenon: Bud 2013.

172 "a tendency to act on inert matter" Bergson 1911, p. 96.

172 A thousand people showed up: McGrath 2013.

173 life's ether: Clément 2015.

173 "In the laboratory": Quoted in Needham 1925, p. 38.

173 Albert Szent-Györgyi: Szent-Györgyi 1963; Bradford 1987; Moss 1988; Robinson 1988; Mommaerts 1992; Rall 2018; "The Albert Szent-Györgyi Papers" n.d.

175 ATP: Engelhardt and Ljubimowa 1939; Schlenk 1987; Maruyama 1991.

179 a series of lectures: Szent-Györgyi 1948.

179 "On the whole what we call life": Quoted in Czapek 1911, p. 63.

179 "the play of water": Quoted in Robinson 1988, p. 217.

179 "What we call 'life'": Quoted in Szent-Györgyi 1948, p. 9.

180 grew famous for his parties: Moss 1988.

180 "subtle reactivity and flexibility": Quoted ibid., p. 243.

181 a cure for cancer: Szent-Györgyi 1977.

181 "I felt, with pain": Quoted in Robinson 1988, p. 230.

181 "I moved from anatomy": Quoted in Szent-Györgyi 1972, p. xxiv.

Scripts

182 For Delbrück, see Delbrück 1970; Harding 1978; Kay 1985; Symonds 1988; McKaughan 2005; Sloan and Fogel 2011; Strauss 2017.

183 "He talked about that a lot": Quoted in Harding 1978.

184 "to jointly consider": Quoted in Sloan and Fogel 2011, p. 61.

185 "We find ourselves fairly gasping": Quoted in Wilson 1923, p. 14.

186 "the basis of life": Quoted in Muller 1929, p. 879.

186 "black market research": Quoted in Harding 1978.

187 Erwin Schrödinger: Kilmister 1987; Phillips 2020.

189 "the fundamental difference": Quoted in Yoxen 1979, p. 33.

189 the nature of life: Schrödinger 2012.

191 Francis Crick: See Crick 1988; Olby 2008; Aicardi 2016.

191 "serve as easy refuge": Quoted in Crick 1988, p. 11.

191 "the dullest problem imaginable": Ibid., p. 13.

192 "the borderline between": Ibid., p. 11.

192 "When it explained": Quoted in Lewis 1947, p. 49.

194 "Now we believe that": Quoted in Tamura 2016, p. 36.

195 "simply smells right": Quoted in "Clue to Chemistry of Heredity Found" 1953, p. 17.

195 "You will see that": Quoted in Cobb 2015, p. 113.

195 one of the most important photographs: Chadarevian 2003.

196 "It is, in a sense": Quoted in Olby 2009, p. 301.

197 "Is Vitalism Dead?": Crick 1966; Hein 1972; Bud 2013; Aicardi 2016.

198 "flogging a dead horse": Quoted in Waddington 1967, p. 202.

198 Sir John Eccles: Eccles 1967.

198 "convince a nineteenth century vitalist": Quoted in Kirschner, Gerhart, and Mitchison 2000, p. 79.

199 "The system must be able to replicate": Quoted in Crick 1982.

199 "We're talking about the search for life": Quoted in Zimmer 2007.

200 *"Life is a self-sustained chemical system"*: Quoted in Joyce 1994, p. xi.

Part Four: Return to the Borderland

Half Life

203 "Mr. Burke": Quoted in Campos 2015, p. 77.

203 anthropause: Rutz et al. 2020.

204 For Covid-19, see Mortensen 2020 and Zimmer 2021.

206 couldn't find bacteria or fungi: Bos 1999; López-García and Moreira 2012.

207 "When one is asked": Quoted in Pirie 1937.

207 crucial features: Pierpont 1999.

209 "According to the working definition": Quoted in Mullen 2013.

210 "Whereas the dream of a normal cell": Quoted in Forterre 2016, p. 104.

210 "alien to logic": Quoted in López-García and Moreira 2012, p. 394.

210 a liter of seawater: Breitbart et al. 2018.

210 a spoonful of dirt: Pratama and Van Elsas 2018.

210 The diversity of viruses: Dion, Oechslin, and Moineau 2020.

211 phages in the ocean: Moniruzzaman et al. 2020.

212 "The life span of erythrocytes": Quoted in Föller, Huber, and Lang 2008, p. 661.

213 "Not a single male has been found": Quoted in Hubbs and Hubbs 1932, p. 629.

215 sexual parasites: Lampert and Schartle 2008; Laskowski et al. 2019.

Data Needed for a Blueprint

216 For Deamer's work, see Deamer 2011; Deamer 2012b; Deamer 2016; Damer 2019; Deamer, Damer, and Kompanichenko 2019; Kompanichenko 2019.

217 "It is mere rubbish thinking": Quoted in Peretó, Bada, and Lazcano 2009, p. 396.

218 to his friends he confided: Strick 2009.

218 "The chief defect": Quoted in Bölsche and McCabe 1906, p. 143.

219 generations of fruit flies: Fry 2000; Mesler and Cleaves II 2015.

219 Alexander Oparin: Broda 1980; Lazcano 2016.

219 "The numerous attempts to discover": Quoted in Oparin 1938, p. 246.

220 "In their delicacy, complexity": Quoted in Oparin 1924, p. 9.

220 disappointing reception: Miller, Schopf, and Lazcano 1997.

221 J. B. S. Haldane: Tirard 2017; Subramanian 2020.

221 "We may, I think": Quoted in Haldane 1929, p. 7.

222 messy waste of time: Lazcano and Bada 2003.

223 "We then designed": Quoted in Miller 1974, p. 232.

224 "Data Needed for a Blueprint . . .": Quoted in Haldane 1965.

224 "The initial organism may have consisted": Quoted in Porcar and Peretó 2018.

224 The idea was potent: Other examples of RNA life theories include Orgel 1968.

226 an abundance of liposomes: Deamer and Bangham 1976.

227 Will Hargreaves: Hargreaves, Mulvihill, and Deamer 1977.

228 formed in space and then fell to Earth: Deamer 2012c; Deamer 2017b.

228 He had made liposomes: Deamer 1985.

230 "The channel must be": Deamer, Akeson, and Branton 2016, p. 518.

231 publish a paper in 1996: Kasianowicz et al. 1996.

231 big bases led to big drops: Akeson et al. 1999.

231 I first met David Deamer: Zimmer 1995.

232 "the RNA World": Quoted in Gilbert 1986, p. 618.

233 this ancient chemistry: Deamer and Barchfeld 1982; Chakrabarti et al. 1994.

235 these vents harbored life: Brazil 2017; Deamer 2017a.

235 the rise of genes, metabolism, and cells: Baross and Hoffman 1985.

236 packed with active volcanoes: Kompanichenko, Poturay, and Shlufman 2015.

237 his powder of life: Deamer 2011.

238 On a trip to New Zealand: Milshteyn et al. 2018.

238 giving it an atmosphere: Deamer 2019.

239 "We've made an RNA-like molecule": Rajamani et al. 2008, p. 73.

239 an orderly arrangement: Deamer 2012a.

239 a Lipid World: Paleos 2015.

242 the 2015 West Africa Ebola outbreak: Quick et al. 2016.

242 identified new insect species: Srivathsan et al. 2019.

242 Kate Adamala: Adamala et al. 2017; Adamala 2019; Gaut et al. 2019.

243 gave Deamer more evidence: Damer and Deamer 2015; Damer et al. 2016; Damer and Deamer 2020.

245 the fossil record: Van Kranendonk, Deamer, and Djokic 2017; Javaux 2019.

245 He still had many opponents: Boyce, Coleman, and Russell 1983; Macleod et al. 1994; Russell 2019.

247 a primitive metabolism could grow: Duval et al. 2019.

247 vitalism: Ibid., p. 10.

247 "utterly irrelevant and misleading": Quoted in Branscomb and Russell 2018a; Branscomb and Russell 2018b.

248 a drug made of RNA: Setten, Rossi, and Han 2019.

248 transthyretin amyloidosis: Lasser et al. 2018.

No Obvious Bushes

250 JPL: Overviews of astrobiology include Dick and Strick 2004 and Kolb 2019.

252 "The fact is that nothing": Quoted in Horowitz 1966, p. 789.

254 home to multicellular organisms: Sagan and Lederberg 1976.

254 "There certainly are no features": Quoted in "Viking I Lands on Mars" 1976.

254 "We can no longer be confident": Quoted in Dick and Strick 2004.

255 Allan Hills 84001: Swartz 1996.

256 it fell to Antarctica: Cavalazzi and Westall 2019.

257 accepted their paper: McKay et al. 1996.

257 "If this discovery is confirmed": Quoted in Clinton 1996.

258 Charles Choi: Choi 2016.

258 "the study of the living universe": Quoted in Dick and Strick 2004.

259 needs liquid water to survive: Kopparapu, Wolf, and Meadows 2019; Shahar et al. 2019.

259 the Great Silence: Ćirković 2018.

260 conditions suitable for life: Hendrix et al. 2019.

261 organic compounds: Postberg et al. 2018.

261 living things in the ocean of Enceladus: Choblet et al. 2017; Kahana, Schmitt-Kopplin, and Lancet 2019.

262 use a nanopore sequencer: Benner 2017; Carr et al. 2020.

264 building her own chimneys: Barge and White 2017.

264 amino acids: Barge et al. 2019.

265 "Since we cannot distinguish": Quoted in Clément 2015.

265 the microbe could grow: Taubner et al. 2018.

266 hover in the borderland: Kahana, Schmitt-Kopplin, and Lancet 2019.

Four Blue Droplets

267 a strangely similar experiment: For the work of Cronin and his colleagues on assembly theory and active matter, see Barge et al. 2015; Cronin, Mehr, and Granda 2018; Doran et al. 2017; Doran, Abul-Haija, and Cronin 2019; Grizou et al. 2019; Grizou et al. 2020; Gromski, Granda, and Cronin 2019; Marshall et al. 2019; Marshall et al. 2020; Miras et al. 2019; Parrilla-Gutierrez et al. 2017; Points et al. 2018; Surman et al. 2019; Walker et al. 2018.

267 Fritz Heider and Marianne Simmel: Heider and Simmel 1944; Scholl and Tremoulet 2000.

270 It narrowed the possibilities: Luisi 1998.

270 somewhere else in some other manner: Cleland 2019b.

270 hundreds of new definitions: The definitions listed here, unless noted below, are from Kolb 2019.

270 *Life is an expected*: Vitas and Dobovišek 2019.

271 *The existence of the dynamically ordered*: Cornish-Bowden and Cárdenas 2020.

271 *Life is a monophyletic clade*: Mariscal and Doolittle 2018.

271 "It is commonly said": Quoted in Westall and Brack 2018, p. 49.

271 Radu Popa: Popa 2004.

272 "Their value does not depend": Quoted in Bich and Green 2018, p. 3933.

273 "Any experiment conducted without": Quoted in Smith 2018, p. 84.

273 *self-reproduction with variations*: Trifonov 2011.

273 "A computer virus performs": Quoted in Meierhenrich 2012, p. 641.

273 Ludwig Wittgenstein: Abbott 2019.

276 space and time: Cleland 1984.

276 the logic of computer programs: Cleland 1993.

278 "The martian-life hypothesis": Quoted in Cleland 1997, p. 20.

279 *The Quest for a Universal Theory of Life*: Cleland 2019a.

279 "Definitions are not the proper tools": Quoted in Cleland 2019b, p. 722.

280 Leonardo da Vinci: Quoted in Cleland 2019a, p. 50.

282 They've created theories: In addition to autocatalytic sets and assembly theory, there are a number of other projects underway. See, for example, England 2020 and Palacios et al. 2020.

283 Robert Rosen and Francisco Varela: Cornish-Bowden and Cárdenas 2020.

283 compressed descriptions: Walker 2018.

283 the essential conditions: Letelier, Cárdenas, and Cornish-Bowden 2011.

283 Stuart Kauffman: Hordijk 2019; Kauffman 2019; Levy 1992.

284 Oil is the product of catalysts: Johns 1979.

285 sustain themselves: Mariscal et al. 2019.

285 Reza Ghadiri: Ashkenasy et al. 2004.

286 a mature theory of life: Hordijk, Shichor, and Ashkenasy 2018; Xavier et al. 2020.

286 self-sustaining chemistry: Hordijk, Steel, and Kauffman 2019.

286 Onnes: Rogalla et al. 2011.

287 Einstein: Schmalian 2010.

289 fifteen steps: Marshall et al. 2020.

289 information: Walker and Davies 2012; Walker, Kim, and Davies 2016; Walker 2017; Davies 2019; Hesp et al. 2019; Palacios et al. 2020.

BIBLIOGRAPHY

Abbott, J. 2019. "Definitions of Life and the Transition from Non-Living to Living." Departmental presentation, Lund University.

Abrams, Marc D. 1998. "The Red Maple Paradox." *BioScience* 48:355–64.

Ackerknecht, Erwin H. 1968. "Death in the History of Medicine." *Bulletin of the History of Medicine* 42:19–23.

Adamala, Katarzyna P., Daniel A. Martin-Alarcon, Katriona R. Guthrie-Honea, and Edward S. Boyden. 2017. "Engineering Genetic Circuit Interactions Within and Between Synthetic Minimal Cells." *Nature Chemistry* 9:431–39.

Adamala, Kate. 2019. "Biology on Sample Size of More Than One." *The 2019 Conference on Artificial Life*. doi:10.1162/isal_a_00124.

Adamatzky, Andrew. 2016. *Advances in Physarum Machines: Sensing and Computing with Slime Mould*. Cham, Switzerland: Springer International Publishing.

Adelman, Juliana. 2007. "Eozoön: Debunking the Dawn Animal." *Endeavour* 31:94–8.

Adolph, Edward F. 1961. "Early Concepts of Physiological Regulations." *Physiological Reviews* 41:737–70.

Aguilar, Pablo S., Mary K. Baylies, Andre Fleissner, Laura Helming, Naokazu Inoue, Benjamin Podbilewicz, Hongmei Wang et al. 2013. "Genetic Basis of Cell-Cell Fusion Mechanisms." *Trends in Genetics* 29:427–37.

Aicardi, Christine. 2016. "Francis Crick, Cross-Worlds Influencer: A Narrative Model to Historicize Big Bioscience." *Studies in History and Philosophy of Science Part C* 55:83–95.

Akeson, Mark, Daniel Branton, John J. Kasianowicz, Eric Brandin, and David W. Deamer. 1999. "Microsecond Time-Scale Discrimination Among Polycytidylic Acid, Polyadenylic Acid, and Polyuridylic Acid as Homopolymers or as Segments Within Single RNA Molecules." *Biophysical Journal* 77: 3227–33.

"The Albert Szent-Györgyi Papers." National Library of Medicine. https://profiles.nlm.nih.gov/spotlight/wg/ (accessed September 2, 2019).

Anderson, James R. 2018. "Chimpanzees and Death." *Philosophical Transactions of the Royal Society B* 373. doi:10.1098/rstb.2017.0257.

Andrew, Audra L., Blair W. Perry, Daren C. Card, Drew R. Schield, Robert P. Ruggiero, Suzanne E. McGaugh, Amit Choudhary et al. 2017. "Growth and Stress Response Mechanisms Underlying Post-Feeding Regenerative Organ Growth in the Burmese Python." *BMC Genomics* 18. doi:10.1186/s12864-017-3743-1.

Andrew, Audra L., Daren C. Card, Robert P. Ruggiero, Drew R. Schield, Richard H. Adams, David D. Pollock, Stephen M. Secor et al. 2015. "Rapid Changes in Gene Expression Direct Rapid Shifts in Intestinal Form and Function in the Burmese Python After Feeding." *Physiological Genomics* 47:147–57.

Ashkenasy, Gonen, Reshma Jagasia, Maneesh Yadav, and M. R. Ghadiri. 2004. "Design of a Directed Molecular Network." *Proceedings of the National Academy of Sciences* 101:10872–7.

Athenaeum, September 12, 1869, p. 339.

Auteri, Giorgia G., and L. L. Knowles. 2020. "Decimated Little Brown Bats Show Potential for Adaptive Change." *Scientific Reports* 10. doi:10.1038/s41598-020-59797-4.

Aviv, Rachel. 2018. "What Does It Mean to Die?" *New Yorker*, February 5. https://www.newyorker.com/magazine/2018/02/05/what-does-it-mean-to-die (accessed June 8, 2020).

Badash, Lawrence. 1978. "Radium, Radioactivity, and the Popularity of Scientific Discovery." *Proceedings of the American Philosophical Society* 122:145–54.

Bains, William. 2014. "What Do We Think Life Is? A Simple Illustration and Its Consequences." *International Journal of Astrobiology* 13:101–11.

Baker, Henry. 1743. *An Attempt Towards a Natural History of the Polype: In a Letter to Martin Folkes*. London: R. Dodsley.

Baker, Henry. 1764. *Employment for the Microscope: In Two Parts*. London: R. & J. Dodsley.

Baker, John R. 1949. "The Cell-Theory: A Restatement, History, and Critique." *Quarterly Journal of Microscopical Science* 90:87–108.

Baker, John R. 1952. *Abraham Trembley of Geneva: Scientist and Philosopher, 1710–1784*. London: Edward Arnold.

Baker, John R. 2008. "Trembley, Abraham." In *Complete Dictionary of Scientific Biography*. Edited by Charles C. Gillispie. New York: Scribner.

Ball, Philip. 2019. *How to Grow a Human: Adventures in How We Are Made and Who We Are*. Chicago: University of Chicago Press.

Bandouchova, Hana, Tomáš Bartonička, Hana Berkova, Jiri Brichta, Tomasz Kokurewicz, Veronika Kovacova, Petr Linhart et al. 2018. "Alterations in the Health of Hibernating Bats Under Pathogen Pressure." *Scientific Reports* 8. doi:10.1038/s41598-018-24461-5.

Barge, Laura M., and Lauren M. White. 2017. "Experimentally Testing Hydrothermal Vent Origin of Life on Enceladus and Other Icy/Ocean Worlds." *Astrobiology* 17:820–33.

Barge, Laura M., Erika Flores, Marc M. Baum, David G. VanderVelde, and Michael J. Russell. 2019. "Redox and pH Gradients Drive Amino Acid Synthesis in Iron Oxyhydroxide Mineral Systems." *Proceedings of the National Academy of Sciences* 116:4828–33.

Barge, Laura M., Silvana S. S. Cardoso, Julyan H. E. Cartwright, Geoffrey J. T. Cooper, Leroy Cronin, Anne De Wit, Ivria J. Doloboff et al. 2015. "From Chemical Gardens to Chemobrionics." *Chemical Reviews* 115:8652–703.

Barnett, James A., and Frieder W. Lichtenthaler. 2001. "A History of Research on Yeasts 3: Emil Fischer, Eduard Buchner and Their Contemporaries, 1880–1900." *Yeast* 18:363–88.

Baross, John A., and Sarah E. Hoffman. 1985. "Submarine Hydrothermal Vents and Associated Gradient Environments as Sites for the Origin and Evolution of Life." *Origins of Life and Evolution of the Biosphere* 15:327–45.

Beale, Lionel S. 1870. *Protoplasm: Or, Life, Force, and Matter*. London: J. Churchill.

Beck, Curt W. 1960. "Georg Ernst Stahl, 1660–1734." *Journal of Chemical Education* 37. doi:10.1021/ed037p506.

Beecher, Henry K. 1968. "A Definition of Irreversible Coma: Report of the Ad Hoc Committee of the Harvard Medical School to Examine the Definition of Brain Death." *Journal of the American Medical Association* 205:337–40.

Benner, Steven A. 2017. "Detecting Darwinism from Molecules in the Enceladus Plumes, Jupiter's Moons, and Other Planetary Water Lagoons." *Astrobiology* 17:840–51.

Bergson, Henri. 1911. *Creative Evolution*. New York: Henry Holt.

Bernal, John D. 1949. "The Physical Basis of Life." *Proceedings of the Physical Society Section B* 62:597–618.

Bernat, James L. 2019. "Refinements in the Organism as a Whole Rationale for Brain Death." *Linacre Quarterly* 86:347–58.

Bernat, James L., and Anne L. D. Ave. 2019. "Aligning the Criterion and Tests for Brain Death." *Cambridge Quarterly of Healthcare Ethics* 28:635–41.

Bernat, James L., Charles M. Culver, and Bernard Gert. 1981. "On the Definition and Criterion of Death." *Annals of Internal Medicine* 94:389–94.

Bernier, Chad R., Anton S. Petrov, Nicholas A. Kovacs, Petar I. Penev, and Loren D. Williams. 2018. "Translation: The Universal Structural Core of Life." *Molecular Biology and Evolution* 35:2065–76.

Berrien, Hank. 2017. "Shapiro Rips Wendy Davis for Claiming Life Beginning at Conception Is 'Absurd.'" *The Daily Wire*, April 30. https://www.dailywire.com/news/shapiro-rips-wendy-davis-claiming-life-beginning-hank-berrien (accessed June 8, 2020).

Berrios, Germán E., and Rogelio Luque. 1995. "Cotard's Delusion or Syndrome?: A Conceptual History." *Comprehensive Psychiatry* 36:218–23.

Berrios, Germán E., and Rogelio Luque. 1999. "Cotard's 'On Hypochondriacal Delusions in a Severe Form of Anxious Melancholia.'" *History of Psychiatry* 10:269–78.

Bich, Leonardo, and Sara Green. 2018. "Is Defining Life Pointless? Operational Definitions at the Frontiers of Biology." *Synthese* 195:3919–46.

Bichat, Xavier. 1815. *Physiological Researches on Life and Death*. London: Longman.

Blackshaw, Bruce P., and Daniel Rodger. 2019. "The Problem of Spontaneous Abortion: Is the Pro-Life Position Morally Monstrous?" *New Bioethics* 25:103–20.

Blackstone, William. 2016. *The Oxford Edition of Blackstone's: Commentaries on the Laws of England*. Oxford: Oxford University Press.

Boback, Scott M., Katelyn J. McCann, Kevin A. Wood, Patrick M. McNeal, Emmett L. Blankenship, and Charles F. Zwemer. 2015. "Snake Constriction Rapidly Induces Circulatory Arrest in Rats." *Journal of Experimental Biology* 218:2279–88.

Boerma, David B., Kenneth S. Breuer, Tim L. Treskatis, and Sharon M. Swartz. 2019. "Wings as Inertial Appendages: How Bats Recover from Aerial Stumbles." *Journal of Experimental Biology* 222. doi:10.1242/jeb.204255.

Bölsche, Wilhelm, and Joseph McCabe. 1906. *Haeckel, His Life and Work.* London: T. F. Unwin.

Bond, George D. 1980. "Theravada Buddhism's Meditations on Death and the Symbolism of Initiatory Death." *History of Religions* 19:237–58.

Bondeson, Jan. 2001. *Buried Alive: The Terrifying History of Our Most Primal Fear*. New York: Norton.

Bos, Lute. 1999. "Beijerinck's Work on Tobacco Mosaic Virus: Historical Context and Legacy." *Philosophical Transactions of the Royal Society B* 354:675–85.

Boussard, Aurèle, Julie Delescluse, Alfonso Pérez-Escudero, and Audrey Dussutour. 2019. "Memory Inception and Preservation in Slime Moulds: The Quest for a Common Mechanism." *Philosophical Transactions of the Royal Society B* 374. doi:10.1098/rstb.2018.0368.

Boyce, Adrian J., M. L. Coleman, and Michael Russell. 1983. "Formation of Fossil Hydrothermal Chimneys and Mounds from Silvermines, Ireland." *Nature* 306:545–50.

Boyles, Justin G., Esmarie Boyles, R. K. Dunlap, Scott A. Johnson, and Virgil Brack Jr. 2017. "Long-Term Microclimate Measurements Add Further Evidence That There Is No 'Optimal' Temperature for Bat Hibernation." *Mammalian Biology* 86:9–16.

Boyles, Justin G., Joseph S. Johnson, Anna Blomberg, and Thomas M. Lilley. 2019. "Optimal Hibernation Theory." *Mammal Review* 50:91–100.

Bradford, H. F. 1987. "A Scientific Odyssey: An Appreciation of Albert Szent-Györgyi." *Trends in Biochemical Sciences* 12:344–47.

Branscomb, Elbert, and Michael J. Russell. 2018a. "Frankenstein or a Submarine Alkaline Vent: Who Is Responsible for Abiogenesis?: Part 1: What Is Life—That It Might Create Itself?" *BioEssays* 40. doi:10.1002/bies.201700179.

Branscomb, Elbert, and Michael J. Russell. 2018b. "Frankenstein or a Submarine Alkaline Vent: Who Is Responsible for Abiogenesis?: Part 2: As Life Is Now, So It Must Have Been in the Beginning." *BioEssays* 40. doi:10.1002/bies.201700182.

Brazil, Rachel. 2017. "Hydrothermal Vents and the Origins of Life." *Chemistry World*, April 16. https://www.chemistryworld.com/features/hydrothermal-vents-and-the-origins-of-life/3007088.article (accessed June 8, 2020).

Breitbart, Mya, Chelsea Bonnain, Kema Malki, and Natalie A. Sawaya. 2018. "Phage Puppet Masters of the Marine Microbial Realm." *Nature Microbiology* 3:754–66.

Brewer, E. N., Susumu Kuraishi, Joseph C. Garver, and Frank M. Strong. 1964. "Mass Culture of a Slime Mold, *Physarum polycephalum*." *Journal of Applied Microbiology* 12:161–64.

Broda, Engelbert. 1980. "Alexander Ivanovich Oparin (1894–1980)." *Trends in Biochemical Sciences* 5:IV–V.

Buchanan, John Y. 1876. "Preliminary Report to Professor Wyville Thomson, F.R.S., Director of the Civilian Scientific Staff, on Work (Chemical and Geological) Done on Board H.M.S. 'Challenger.'" *Proceedings of the Royal Society* 24:593–623.

Buchner, Eduard. 1907. "Nobel Lecture: Cell-Free Fermentation" *The Nobel Prize*, December 11. https://www.nobelprize.org/prizes/chemistry/1907/buchner/lecture/ (accessed June 8, 2020).

Bud, Robert. 2013. "Life, DNA and the Model." *British Journal for the History of Science* 46:311–34.

Burke, John B. (n.d.). MS Archives of the Royal Literary Fund. *Nineteenth Century Collections Online*.

Burke, John B. 1903. "The Radio-Activity of Matter." *Monthly Review* 13:115–31.

Burke, John B. 1905a. "On the Spontaneous Action of Radio-Active Bodies on Gelatin Media." *Nature* 72:78–79.

Burke, John B. 1905b. "The Origin of Life." *Fortnightly Review* 78:389–402.

Burke, John B. 1906. *The Origin of Life: Its Physical Basis and Definition*. London: Chapman & Hall.

Burke, John B. 1931a. *The Emergence of Life*. London: Oxford University Press.

Burke, John B. 1931b. *The Mystery of Life*. London: Elkin Mathews & Marrot.

Burkhardt, Frederick, Duncan M. Porter, Sheila A. Dean, Jonathan R. Topham, and Sarah Wilmot. 1999. *The Correspondence of Charles Darwin: Volume 11, 1863*. Cambridge, UK: Cambridge University Press.

Burns, Russell M., and Barbara H. Honkala. 1990. "Silvics of North America." In *Agriculture Handbook 654*. Washington, D.C.: U.S. Department of Agriculture.

"The Cambridge Radiobes." *New York Tribune*, July 2, 1905, p. 11.

Campbell, George G. 1877. *Log-Letters from "The Challenger."* London: Macmillan.

Campbell, Norman R. 1906. "Sensationalism and Science." *National Review* 48:89–99.

Campos, Luis. 2006. "Radium and the Secret of Life." PhD dissertation, Harvard University.

Campos, Luis. 2007. "The Birth of Living Radium." *Representations* 97:1–27.

Campos, Luis. 2015. *Radium and the Secret of Life*. Chicago: University of Chicago Press.

Cannone, Nicoletta, T. Corinti, Francesco Malfasi, P. Gerola, Alberto Vianelli, Isabella Vanetti, S. Zaccara et al. 2017. "Moss Survival Through *in situ* Cryptobiosis After Six Centuries of Glacier Burial." *Scientific Reports* 7. doi:10.1038/s41598-017-04848-6.

Caramazza, Alfonso, and Jennifer R. Shelton. 1998. "Domain-Specific Knowledge Systems in the Brain: The Animate-Inanimate Distinction." *Journal of Cognitive Neuroscience* 10:1–34.

Carpenter, William B. 1864. "On the Structure and Affinities of *Eozoon canadense*." *Proceedings of the Royal Society* 13:545–49.

Carr, Christopher E., Noelle C. Bryan, Kendall N. Saboda, Srinivasa A. Bhattaru, Gary Ruvkun, and Maria T. Zuber. 2020. "Nanopore Sequencing at Mars, Europa and Microgravity Conditions." doi:10.1101/2020.01.09.899716.

Cavalazzi, Barbara, and Frances Westall. 2019. *Biosignatures for Astrobiology*. Cham, Switzerland: Springer International Publishing.

Cavendish Library. 1910. *A History of the Cavendish Laboratory 1871–1910*. London: Longmans, Green & Co.

Chadarevian, Soraya de. 2003. "Portrait of a Discovery: Watson, Crick, and the Double Helix." *Isis* 94:90–105.

Chakrabarti, Ajoy C., Ronald R. Breaker, Gerald F. Joyce, and David W. Deamer. 1994. "Production of RNA by a Polymerase Protein Encapsulated Within Phospholipid Vesicles." *Journal of Molecular Evolution* 39:555–59.

Chatterjee, Seshadri S., and Sayantanava Mitra. 2015. "'I Do Not Exist'—Cotard Syndrome in Insular Cortex Atrophy." *Biological Psychiatry* 77:e52–53.

Choblet, Gaël, Gabriel Tobie, Christophe Sotin, Marie Běhounková, Ondřej Čadek, Frank Postberg, and Ondřej Souček. 2017. "Powering Prolonged Hydrothermal Activity Inside Enceladus." *Nature Astronomy* 1:841–47.

Choi, Charles Q. 2016. "Mars Life? 20 Years Later, Debate over Meteorite Continues." Space.com, August 10. https://www.space.com/33690-allen-hills-mars-meteorite-alien-life-20-years.html (accessed July 25, 2020).

Cipriani, Gabriele, Angelo Nuti, Sabrina Danti, Lucia Picchi, and Mario Di Fiorino. 2019. "'I Am Dead': Cotard Syndrome and Dementia." *International Journal of Psychiatry in Clinical Practice* 23:149–56.

Ćirković, Milan M. 2018. *The Great Silence: Science and Philosophy of Fermi's Paradox*. New York: Oxford University Press.

"City Chatter." *Sunday Times*, June 25, 1905, p. 3.

Clark, Jim, and Edward F. Haskins. 2010. "Reproductive Systems in the Myxomycetes: A Review." *Mycosphere* 1:337–53.

Clegg, James S. 2001. "Cryptobiosis—A Peculiar State of Biological Organization." *Comparative Biochemistry and Physiology Part B* 128:613–24.

Cleland, Carol E. 1984. "Space: An Abstract System of Non-Supervenient Relations." *Philosophical Studies: An International Journal for Philosophy in the Analytic Tradition* 46:19–40.

Cleland, Carol E. 1993. "Is the Church-Turing Thesis True?" *Minds and Machines* 3:283–312.

Cleland, Carol E. 1997. "Standards of Evidence: How High for Ancient Life on Mars?" *Planetary Report* 17:20–21.

Cleland, Carol E. 2019a. *The Quest for a Universal Theory of Life: Searching for Life as We Don't Know It*. New York: Cambridge University Press.

Cleland, Carol E. 2019b. "Moving Beyond Definitions in the Search for Extraterrestrial Life." *Astrobiology* 19:722–29.

Clément, Raphaël. 2015. "Stéphane Leduc and the Vital Exception in the Life Sciences." arXiv:1512.03660.

Clinton, William J. 1996. "President Clinton Statement Regarding Mars Meteorite Discovery." Jet Propulsion Laboratory, August 7. https://www2.jpl.nasa.gov/snc/clinton.html (accessed June 8, 2020).

"Clue to Chemistry of Heredity Found." *New York Times*, June 13, 1953, p. 17.

"A Clue to the Beginning of Life on the Earth." *World's Work*, November 1905, 11:6813–14.

Cobb, Matthew. 2015. *Life's Greatest Secret: The Race to Crack the Genetic Code*. New York: Basic Books.

Connolly, Andrew C., Long Sha, J. S. Guntupalli, Nikolaas Oosterhof, Yaroslav O. Halchenko, Samuel A. Nastase, Matteo V. Di Oleggio Castello et al. 2016. "How the Human Brain Represents Perceived Dangerousness or 'Predacity' of Animals." *Journal of Neuroscience* 36:5373–84.

Cooper, Vaughn S., Taylor M. Warren, Abigail M. Matela, Michael Handwork, and Shani Scarponi. 2019. "EvolvingSTEM: A Microbial Evolution-in-Action Curriculum That Enhances Learning of Evolutionary Biology and Biotechnology." *Evolution: Education and Outreach* 12. doi:10.1186/s12052-019-0103-4.

Cornish-Bowden, Athel, and María L. Cárdenas. 2020. "Contrasting Theories of Life: Historical Context, Current Theories. In Search of an Ideal Theory." *Biosystems* 188. doi:10.1016/j.biosystems.2019.104063.

Crick, Francis. 1966. *Of Molecules and Men: A Volume in The John Danz Lectures Series*. Seattle: University of Washington Press.

Crick, Francis. 1982. *Life Itself: Its Origin and Nature*. New York: Simon & Schuster.

Crick, Francis. 1988. *What Mad Pursuit: A Personal View of Scientific Discovery*. New York: Basic Books.

Crippen, Tawni L., Mark E. Benbow, and Jennifer L. Pechal. 2015. "Microbial Interactions During Carrion Decomposition." In *Carrion Ecology, Evolution, and Their Applications*. Edited by Mark E. Benbow, Jeffery K. Tomberlin, and Aaron M. Tarone. Boca Raton, FL: CRC Press.

Cronin, Leroy, S. H. M. Mehr, and Jarosław M. Granda. 2018. "Catalyst: The Metaphysics of Chemical Reactivity." *Chem* 4:1759–61.

Cryan, Paul M., Carol U. Meteyer, Justin Boyles, and David S. Blehert. 2010. "Wing Pathology of White-Nose Syndrome in Bats Suggests Life-Threatening Disruption of Physiology." *BMC Biology* 8. doi:10.1186/1741-7007-8-135.

Cugola, Fernanda R., Isabella R. Fernandes, Fabiele B. Russo, Beatriz C. Freitas, João L. M. Dias, Katia P. Guimarães, Cecília Benazzato et al. 2016. "The Brazilian Zika Virus Strain Causes Birth Defects in Experimental Models." *Nature* 534:267–71.

Cunningham, Andrew. 2016. *The Anatomist Anatomis'd: An Experiment Discipline in Enlightenment Europe*. London: Routledge.

Czapek, Friedrich. 1911. *Chemical Phenomena in Life*. London: Harper & Bros.

Damer, Bruce, and David Deamer. 2015. "Coupled Phases and Combinatorial Selection in Fluctuating Hydrothermal Pools: A Scenario to Guide Experimental Approaches to the Origin of Cellular Life." *Life* 5:872–87.

Damer, Bruce, and David Deamer. 2020. "The Hot Spring Hypothesis for an Origin of Life." *Astrobiology* 20:429–52.

Damer, Bruce, David Deamer, Martin Van Kranendonk, and Malcolm Walter. 2016. "An Origin of Life Through Three Coupled Phases in Cycling Hydrothermal Pools with Distribution and Adaptive Radiation to Marine Stromatolites." In *Proceedings of the 2016 Gordon Research Conference on the Origins of Life*.

Damer, Bruce. 2019. "David Deamer: Five Decades of Research on the Question of How Life Can Begin." *Life* 9. doi:10.3390/life9020036.

Darwin, Charles. 1871. *The Descent of Man, and Selection in Relation to Sex*. New York: D. Appleton.

Davies, Paul C. W. 2019. *The Demon in the Machine: How Hidden Webs of Information Are Solving the Mystery of Life*. Chicago: University of Chicago Press.

Dawson, Virginia P. 1987. *Nature's Enigma: The Problem of the Polyp in the Letters of Bonnet, Trembley and Réaumur*. Philadelphia: American Philosophical Society.

Dawson, Virginia P. 1991. "Regeneration, Parthenogenesis, and the Immutable Order of Nature." *Archives of Natural History* 18:309–21.

De Jong, Piet C. 1976. *Flowering and Sex Expression in Acer L.: A Biosystematic Study*. Wageningen: Veenman.

Deamer, David W. 1985. "Boundary Structures Are Formed by Organic Components of the Murchison Carbonaceous Chondrite." *Nature* 317: 792–94.

Deamer, David W. 1998. "Daniel Branton and Freeze-Fracture Analysis of Membranes." *Trends in Cell Biology* 8:460–62.

Deamer, David W. 2010. "From 'Banghasomes' to Liposomes: A Memoir of Alec Bangham, 1921–2010." *FASEB Journal* 24:1308–10.

Deamer, David W. 2011. "Sabbaticals, Self-Assembly, and Astrobiology." *Astrobiology* 11:493–98.

Deamer, David W. 2012a. "Liquid Crystalline Nanostructures: Organizing Matrices for Non-Enzymatic Nucleic Acid Polymerization." *Chemical Society Reviews* 41:5375–79.

Deamer, David W. 2012b. *First Life: Discovering the Connections Between Stars, Cells, and How Life Began*. Berkeley: University of California Press.

Deamer, David W. 2012c. "Membranes, Murchison, and Mars: An Encapsulated Life in Science." *Astrobiology* 12:616–17.

Deamer, David W. 2016. "Membranes and the Origin of Life: A Century of Conjecture." *Journal of Molecular Evolution* 83:159–68.

Deamer, David W. 2017a. "Conjecture and Hypothesis: The Importance of Reality Checks." *Beilstein Journal of Organic Chemistry* 13:620–24.

Deamer, David W. 2017b. "Darwin's Prescient Guess." *Proceedings of the National Academy of Sciences* 114:11264–65.

Deamer, David W. 2019. *Assembling Life: How Can Life Begin on Earth and Other Habitable Planets?* New York: Oxford University Press.

Deamer, David W., and Alec D. Bangham. 1976. "Large Volume Liposomes by an Ether Vaporization Method." *Biochimica et Biophysica Acta* 443: 629–34.

Deamer, David W., and Daniel Branton. 1967. "Fracture Planes in an Ice-Bilayer Model Membrane System." *Science* 158:655–57.

Deamer, David W., and Gail L. Barchfeld. 1982. "Encapsulation of Macromolecules by Lipid Vesicles Under Simulated Prebiotic Conditions." *Journal of Molecular Evolution* 18:203–6.

Deamer, David W., Bruce Damer, and Vladimir Kompanichenko. 2019. "Hydrothermal Chemistry and the Origin of Cellular Life." *Astrobiology* 19:1523–37.

Deamer, David W., Mark Akeson, and Daniel Branton. 2016. "Three Decades of Nanopore Sequencing." *Nature Biotechnology* 34:518–24.

Deamer, David W., Robert Leonard, Annette Tardieu, and Daniel Branton. 1970. "Lamellar and Hexagonal Lipid Phases Visualized by Freeze-Etching." *Biochimica et Biophysica Acta* 219:47–60.

Debruyne, Hans, Michael Portzky, Frédérique Van Den Eynde, and Kurt Audenaert. 2009. "Cotard's Syndrome: A Review." *Current Psychiatry Reports* 11:197–202.

Delbrück, Max. 1970. "A Physicist's Renewed Look at Biology: Twenty Years Later." *Science* 168:1312–15.

Desmond, Adrian J. 1999. *Huxley: From Devil's Disciple to Evolution's High Priest*. New York: Basic Books.

Devolder, Katrien, and John Harris. 2007. "The Ambiguity of the Embryo: Ethical Inconsistency in the Human Embryonic Stem Cell Debate." *Metaphilosophy* 38:153–69.

Diamond, Jared. 1994. "Dining with the Snakes." *Discover*, January 18. https://www.discovermagazine.com/the-sciences/dining-with-the-snakes (accessed June 8, 2020).

Dick, Steven J., and James E. Strick. 2004. *The Living Universe: NASA and the Development of Astrobiology*. New Brunswick, NJ: Rutgers University Press.

Dieguez, Sebastian. 2018. "Cotard Syndrome." *Frontiers of Neurology and Neuroscience* 42:23–34.

Di Giorgio, Elisa, Marco Lunghi, Francesca Simion, and Giorgio Vallortigara. 2017. "Visual Cues of Motion That Trigger Animacy Perception at Birth: The Case of Self-Propulsion." *Developmental Science* 20. doi:10.1111/desc.12394.

Dion, Moïra B., Frank Oechslin, and Sylvain Moineau. 2020. "Phage Diversity, Genomics and Phylogeny." *Nature Reviews Microbiology* 18:125–38.

Dolan, Chris. 2018. "Jahi McMath Has Died in New Jersey." *Dolan Law Firm*, June 29. https://dolanlawfirm.com/2018/06/jahi-mcmath-has-died-in-new-jersey/ (accessed June 8, 2020).

Doran, David, Marc Rodriguez-Garcia, Rebecca Turk-MacLeod, Geoffrey J. T. Cooper, and Leroy Cronin. 2017. "A Recursive Microfluidic Platform to Explore the Emergence of Chemical Evolution." *Beilstein Journal of Organic Chemistry* 13:1702–9.

Doran, David, Yousef M. Abul-Haija, and LeRoy Cronin. 2019. "Emergence of Function and Selection from Recursively Programmed Polymerisation Reactions in Mineral Environments." *Angewandte Chemie International Edition* 58:11253–56.

Douglas Rudge, W. A. 1906. "The Action of Radium and Certain Other Salts on Gelatin." *Proceedings of the Royal Society A* 78:380–84.

Dussutour, Audrey, Tanya Latty, Madeleine Beekman, and Stephen J. Simpson. 2010. "Amoeboid Organism Solves Complex Nutritional Challenges." *Proceedings of the National Academy of Sciences* 107:4607–11.

Duval, Simon, Frauke Baymann, Barbara Schoepp-Cothenet, Fabienne Trolard, Guilhem Bourrié, Olivier Grauby, Elbert Branscomb et al. 2019. "Fougerite: The Not So Simple Progenitor of the First Cells." *Interface Focus* 9. doi:10.1098/rsfs.2019.0063.

Dyson, Ketaki K. 1978. *A Various Universe: A Study of the Journals and Memoirs of British Men and Women in the Indian Subcontinent, 1765–1856*. New York: Oxford University Press.

Eccles, John C. 1967. "Book Review of 'Of Molecules and Men,' by Frances Crick." *Zygon* 2:281–82.

El Hachem, Hady, Vincent Crepaux, Pascale May-Panloup, Philippe Descamps, Guillaume Legendre, and Pierre-Emmanuel Bouet. 2017. "Recurrent Pregnancy Loss: Current Perspectives." *International Journal of Women's Health* 9:331–45.

Engber, Daniel. 2017. "When the Lab Rat Is a Snake." *New York Times*, May 17. https://www.nytimes.com/2017/05/17/magazine/when-the-lab-rat-is-a-snake .html (accessed June 8, 2020).

Engelhardt, Wladimir A., and Militza N. Ljubimowa. 1939. "Myosine and Adenosinetriphosphatase." *Nature* 144:668–69.

English, Jeremy. 2020. *Every Life Is on Fire: How Thermodynamics Explains the Origins of Living Things*. New York: Basic Books.

Ferguson, Gayle C., Frederic Bertels, and Paul B. Rainey. 2013. "Adaptive Divergence in Experimental Populations of *Pseudomonas fluorescens*. V. Insight into the Niche Specialist Fuzzy Spreader Compels Revision of the Model *Pseudomonas* Radiation." *Genetics* 195:1319–35.

"Filipino Scientist, A." *Filipino*, 1906, 1:5.

Flynn, Kenneth M., Gabrielle Dowell, Thomas M. Johnson, Benjamin J. Koestler, Christopher M. Waters, and Vaughn S. Cooper. 2016. "Evolution of Ecological Diversity in Biofilms of *Pseudomonas aeruginosa* by Altered Cyclic Diguanylate Signaling." *Journal of Bacteriology* 198:2608–18.

Föller, Michael, Stephan M. Huber, and Florian Lang. 2008. "Erythrocyte Programmed Cell Death." *IUBMB Life* 60:661–68.

Forbes, James. 1813. *Oriental Memoirs: Selected and Abridged from a Series of Familiar Letters Written During Seventeen Years Residence in India: Including Observations on Parts of Africa and South America, and a Narrative of Occurrences in Four India Voyages: Illustrated by Engravings from Original Drawings*. London: White, Cochrane & Co.

Forterre, Patrick. 2016. "To Be or Not to Be Alive: How Recent Discoveries Challenge the Traditional Definitions of Viruses and Life." *Studies in History and Philosophy of Science Part C* 59:100–108.

Fox, Robert, and Cynthia McDaniel. 1982. "The Perception of Biological Motion by Human Infants." *Science* 218:486–87.

Fraser, James A., and Joseph Heitman. 2003. "Fungal Mating-Type Loci." *Current Biology* 13:R792–95.

Frixione, Eugenio. 2006. "Albrecht Von Haller (1708–1777)." *Journal of Neurology* 253:265–66.

Frixione, Eugenio. 2007. "Irritable Glue: The Haller-Whytt Controversy on the Mechanism of Muscle Contraction." In *Brain, Mind and Medicine: Essays in Eighteenth-Century Neuroscience*. Edited by Harry Whitaker, C. U. M. Smith, and Stanley Finger. Boston: Springer.

Fry, Iris. 2000. *The Emergence of Life on Earth: A Historical and Scientific Overview*. New Brunswick, NJ: Rutgers University Press.

Gambarotto, Andrea. 2018. *Vital Forces, Teleology and Organization: Philosophy of Nature and the Rise of Biology in Germany*. Cham, Switzerland: Springer International Publishing.

Gao, Chao, Chen Liu, Daniel Schenz, Xuelong Li, Zili Zhang, Marko Jusup, Zhen Wang et al. 2019. "Does Being Multi-Headed Make You Better at Solving Problems? A Survey of *Physarum*-Based Models and Computations." *Physics of Life Reviews* 29:1–26.

Gaut, Nathaniel J., Jose Gomez-Garcia, Joseph M. Heili, Brock Cash, Qiyuan Han, Aaron E. Engelhart, and Katarzyna P. Adamala. 2019. "Differentiation of Pluripotent Synthetic Minimal Cells via Genetic Circuits and Programmable Mating." doi:10.1101/712968.

Geison, Gerald L. 1969. "The Protoplasmic Theory of Life and the Vitalist-Mechanist Debate." *Isis* 60:272–92.

Giakoumelou, Sevi, Nick Wheelhouse, Kate Cuschieri, Gary Entrican, Sarah E. M. Howie, and Andrew W. Horne. 2016. "The Role of Infection in Miscarriage." *Human Reproduction Update* 22:116–33.

Gibson, Susannah. 2015. *Animal, Vegetable, Mineral?: How Eighteenth-Century Science Disrupted the Natural Order*. New York: Oxford University Press.

Gignoux-Wolfsohn, Sarah A., Malin L. Pinsky, Kathleen Kerwin, Carl Herzog, Mackenzie Hall, Alyssa B. Bennett, Nina H. Fefferman et al. 2018. "Genomic Signatures of Evolutionary Rescue in Bats Surviving White-Nose Syndrome." doi:10.1101/470294.

Gilbert, Walter. 1986. "Origin of Life: The RNA World." *Nature* 319. doi:10.1038/319618a0.

Gloag, Erin S., Christopher W. Marshall, Daniel Snyder, Gina R. Lewin, Jacob S. Harris, Alfonso Santos-Lopez, Sarah B. Chaney et al. 2019. "*Pseudomonas aeruginosa* Interstrain Dynamics and Selection of Hyperbiofilm Mutants During a Chronic Infection." *mBio* 10. doi:10.1128/mBio.01698-19.

Gloag, Erin S., Christopher W. Marshall, Daniel Snyder, Gina R. Lewin, Jacob S. Harris, Sarah B. Chaney, Marvin Whiteley et al. 2018. "The *Pseudomonas aeruginosa* Wsp Pathway Undergoes Positive Evolutionary Selection During Chronic Infection." doi:10.1101/456186.

Gonçalves, André, and Dora Biro. 2018. "Comparative Thanatology, an Integrative Approach: Exploring Sensory/Cognitive Aspects of Death Recognition in Vertebrates and Invertebrates." *Philosophical Transactions of the Royal Society B* 373. doi:10.1098/rstb.2017.0263.

Gonçalves, André, and Susana Carvalho. 2019. "Death Among Primates: A Critical Review of Non-Human Primate Interactions Towards Their Dead and Dying." *Biological Reviews* 94. doi:10.1111/brv.12512.

Gottlieb, Alma. 2004. *The Afterlife Is Where We Come From: The Culture of Infancy in West Africa*. Chicago: University of Chicago Press.

Goulon, Maurice, P. Babinet, and N. Simon. 1983. "Brain Death or Coma Dépassé." In *Care of the Critically Ill Patient*. Edited by Jack Tinker and Maurice Rapin. Berlin: Springer-Verlag.

Graber, Raymond E., and William B. Leak. 1992. "Seed Fall in an Old-Growth Northern Hardwood Forest." *U.S. Department of Agriculture*. doi:10.2737/NE-RP-663.

Green, Douglas S. 1980. "The Terminal Velocity and Dispersal of Spinning Samaras." *American Journal of Botany* 67:1218–24.

Greene, D. F., and E. A. Johnson. 1992. "Fruit Abscission in *Acer saccharinum* with Reference to Seed Dispersal." *Canadian Journal of Botany* 70:2277–83.

Greene, D. F., and E. A. Johnson. 1993. "Seed Mass and Dispersal Capacity in Wind-Dispersed Diaspores." *Oikos* 67:69–74.

Grizou, Jonathan, Laurie J. Points, Abhishek Sharma, and Leroy Cronin. 2019. "Exploration of Self-Propelling Droplets Using a Curiosity Driven Robotic Assistant." arXiv:1904.12635.

Grizou, Jonathan, Laurie J. Points, Abhishek Sharma, and Leroy Cronin. 2020. "A Curious Formulation Robot Enables the Discovery of a Novel Protocell Behavior." *Science Advances* 6. doi:10.1126/sciadv.aay4237.

Gromski, Piotr S., Jarosław M. Granda, and Leroy Cronin. 2019. "Universal Chemical Synthesis and Discovery with 'The Chemputer.'" *Trends in Chemistry* 2:4–12.

Guardian, May 25, 1905, p. 6.

Gulliver, George. 1873. "Tears and Care of Monkeys for the Dead." *Nature* 8. doi:10.1038/008103c0.

Haas, David M., Taylor J. Hathaway, and Patrick S. Ramsey. 2019. "Progestogen for Preventing Miscarriage in Women with Recurrent Miscarriage of Unclear Etiology." *Cochrane Database of Systematic Reviews*. doi:10.1002/14651858.CD003511.pub5.

Haase, Catherine G., Nathan W. Fuller, C. R. Hranac, David T. S. Hayman, Liam P. McGuire, Kaleigh J. O. Norquay, Kirk A. Silas et al. 2019. "Incorporating Evaporative Water Loss into Bioenergetic Models of Hibernation to Test for Relative Influence of Host and Pathogen Traits on White-Nose Syndrome." *PLoS One* 14. doi:10.1371/journal.pone.0222311.

Haigh, Elizabeth. 1984. *Xavier Bichat and the Medical Theory of the Eighteenth Century (Medical History, Supplement No. 4)*. London: Wellcome Institute for the History of Medicine.

Haldane, John B. S. 1929. "The Origin of Life." Reprinted in *Origin of Life*. Edited by John D. Bernal. Cleveland, OH: World Publishing Company.

Haldane, John B. S. 1947. *What Is Life?* New York: Boni & Gaer.

Haldane, John B. S. 1965. "Data Needed for a Blueprint of the First Organism." In *The Origins of Prebiological Systems and of their Molecular Matrices*. Edited by Sidney W. Fox. New York: Academic Press.

Hale, William B. 1905. "Has Radium Revealed the Secret of Life?" *New York Times*, July 16, p. 7.

Haller, Albrecht V., and O. Temkin. 1936. *A Dissertation on the Sensible and Irritable Parts of Animals*. Baltimore: Johns Hopkins University Press.

Harding, Carolyn. 1978. "Interview with Max Delbruck." *Caltech Institute Archives*, September 11. https://resolver.caltech.edu/CaltechOH:OH_Delbruck_M (accessed June 8, 2020).

Hargreaves, W. R., Sean J. Mulvihill, and David W. Deamer. 1977. "Synthesis of Phospholipids and Membranes in Prebiotic Conditions." *Nature* 266:78–80.

Hedenström, Anders, and L. C. Johansson. 2015. "Bat Flight: Aerodynamics, Kinematics and Flight Morphology." *Journal of Experimental Biology* 218:653–63.

Heider, Fritz, and Marianne Simmel. 1944. "An Experimental Study of Apparent Behavior." *American Journal of Psychology* 57:243–59.

Hein, Hilde. 1972. "The Endurance of the Mechanism: Vitalism Controversy." *Journal of the History of Biology* 5:159–88.

Hendrix, Amanda R., Terry A. Hurford, Laura M. Barge, Michael T. Bland, Jeff S. Bowman, William Brinckerhoff, Bonnie J. Buratti et al. 2019. "The NASA Roadmap to Ocean Worlds." *Astrobiology* 19:1–27.

Herbst, Charles C., and George R. Johnstone. 1937. "Life History of *Pelagophycus porra*." *Botanical Gazette* 99:339–54.

Hesp, Casper, Maxwell J. D. Ramstead, Axel Constant, Paul Badcock, Michael Kirchhoff, and Karl J. Friston. 2019. "A Multi-Scale View of the Emergent Complexity of Life: A Free-Energy Proposal." In *Evolution, Development, and Complexity: Multiscale Models in Complex Adaptive Systems*. Edited by Georgi Y. Georgiev, John M. Smart, Claudio L. Flores Martinez, and Michael E. Price. Cham, Switzerland: Springer International Publishing.

Hintzsche, Erich. 2008. "Haller, (Victor) Albrecht Von." In *Complete Dictionary of Scientific Biography*. Edited by Charles C. Gillispie. New York: Scribner.

Hintzsche, Erich, and Jörn H. Wolf. 1962. *Albrecht von Hallers Abhandlung über die Wirkung des Opiums auf den menschlichen Körper: übersetzt und erläutert*. Bern: Paul Haupt.

Hoffman, Friedrich. 1971. *Fundamenta medicinae*. Translated by Lester King. London: Macdonald.

Hordijk, Wim. 2019. "A History of Autocatalytic Sets: A Tribute to Stuart Kauffman." *Biological Theory* 14:224–46.

Hordijk, Wim, Mike Steel, and Stuart A. Kauffman. 2019. "Molecular Diversity Required for the Formation of Autocatalytic Sets." *Life* 9:23.

Hordijk, Wim, Shira Shichor, and Gonen Ashkenasy. 2018. "The Influence of Modularity, Seeding, and Product Inhibition on Peptide Autocatalytic Network Dynamics." *ChemPhysChem* 19:2437–44.

Horowitz, Norman H. 1966. "The Search for Extraterrestrial Life." *Science* 151:789–92.

Hostiuc, Sorin, Mugurel C. Rusu, Ionuţ Negoi, Paula Perlea, Bogdan Dorobanţu, and Eduard Drima. 2019. "The Moral Status of Cerebral Organoids." *Regenerative Therapy* 10:118–22.

Houle, Gilles, and Serge Payette. 1991. "Seed Dynamics of *Abies balsamea* and *Acer saccharum* in a Deciduous Forest of Northeastern North America." *American Journal of Botany* 78:895–905.

Hovers, Erella, and Anna Belfer-Cohen. 2013. "Insights into Early Mortuary Practices of *Homo*." In *The Oxford Handbook of the Archaeology of Death and Burial*. Edited by Liv N. Stutz and Sarah Tarlow. Oxford: Oxford University Press.

Huang, Andrew P., and James L. Bernat. 2019. "The Organism as a Whole in an Analysis of Death." *Journal of Medicine and Philosophy* 44:712–31.

Hubbs, Carl L., and Laura C. Hubbs. 1932. "Apparent Parthenogenesis in Nature, in a Form of Fish of Hybrid Origin." *Science* 76:628–30.

Huber, Christian G., and Agorastos. 2012. "We Are All Zombies Anyway: Aggression in Cotard's Syndrome." *Journal of Neuropsychiatry and Clinical Neurosciences* 24. doi:10.1176/appi.neuropsych.11070155.

Hughes, Jeffrey W., and Timothy J. Fahey. 1988. "Seed Dispersal and Colonization in a Disturbed Northern Hardwood Forest." *Bulletin of the Torrey Botanical Club* 115:89–99.

Hunter, Graeme K. 2000. *Vital Forces: The Discovery of the Molecular Basis of Life*. London: Academic Press.

Hussain, Ashiq, Luis R. Saraiva, David M. Ferrero, Gaurav Ahuja, Venkatesh S. Krishna, Stephen D. Liberles, and Sigrun I. Korsching. 2013. "High-Affinity Olfactory Receptor for the Death-Associated Odor Cadaverine." *Proceedings of the National Academy of Sciences* 110:19579–84.

Huxley, Thomas H. 1868. "On Some Organisms Living at Great Depths in the North Atlantic Ocean." *Quarterly Journal of Microscopical Science* 8:203–12.

Huxley, Thomas H. 1869. "On the Physical Basis of Life." *Fortnightly Review* 5:129–45.

Huxley, Thomas H. 1875. "Notes from the 'Challenger.'" *Nature* 12:315–16.

Huxley, Thomas H. 1891. "Biology." In *Encyclopaedia Britannica*. Philadelphia: Maxwell Somerville.

Jarvis, Gavin E. 2016a. "Early Embryo Mortality in Natural Human Reproduction: What the Data Say." *F1000Research* 5. doi:10.12688/f1000research.8937.2.

Jarvis, Gavin E. 2016b. "Estimating Limits for Natural Human Embryo Mortality." *F1000Research* 5. doi:10.12688/f1000research.9479.1.

Javaux, Emmanuelle J. 2019. "Challenges in Evidencing the Earliest Traces of Life." *Nature* 572:451–60.

Johns, William D. 1979. "Clay Mineral Catalysis and Petroleum Generation." *Annual Review of Earth and Planetary Sciences* 7:183–98.

Johnson, Joseph S., Michael R. Scafini, Brent J. Sewall, and Gregory G. Turner. 2016. "Hibernating Bat Species in Pennsylvania Use Colder Winter Habitats Following the Arrival of White-nose Syndrome." In *Conservation and Ecology of Pennsylvania's Bats*. Edited by Calvin M. Butchkoski, DeeAnn M. Reeder, Gregory G. Turner, and Howard P. Whidden. East Stroudsburg, PA: Pennsylvania Academy of Science.

Joyce, Gerald F. 1994. "Foreword." In *Origins of Life: The Central Concepts*. Edited by David W. Deamer and Gail R. Fleischaker. Boston: Jones & Bartlett.

Kahana, Amit, Philippe Schmitt-Kopplin, and Doron Lancet. 2019. "Enceladus: First Observed Primordial Soup Could Arbitrate Origin-of-Life Debate." *Astrobiology* 19:1263–78.

Kasianowicz, John J., Eric Brandin, Daniel Branton, and David W. Deamer. 1996. "Characterization of Individual Polynucleotide Molecules Using a Membrane Channel." *Proceedings of the National Academy of Sciences* 93:13770–73.

Kauffman, Stuart A. 2019. *A World Beyond Physics: The Emergence and Evolution of Life*. Oxford: Oxford University Press.

Kay, Lily E. 1985. "Conceptual Models and Analytical Tools: The Biology of Physicist Max Delbrück." *Journal of the History of Biology* 18:207–46.

Keilin, David. 1959. "The Leeuwenhoek Lecture: The Problem of Anabiosis or Latent Life: History and Current Concept." *Proceedings of the Royal Society B* 150:149–91.

Kilmister, Clive W. 1987. *Schrödinger: Centenary Celebration of a Polymath*. Cambridge, UK: Cambridge University Press.

King-Hele, Desmond. 1998. "The 1997 Wilkins Lecture: Erasmus Darwin, the Lunaticks and Evolution." *Notes and Records* 52:153–80.

King, William, and T. H. Rowney. 1869. "On the So-Called 'Eozoonal' Rock." *Quarterly Journal of the Geological Society* 25:115–18.

Kirschner, Marc, John Gerhart, and Tim Mitchison. 2000. "Molecular 'Vitalism.'" *Cell* 100:79–88.

Koch, Christof. 2019a. *The Feeling of Life Itself: Why Consciousness Is Widespread but Can't Be Computed.* Cambridge, MA: MIT Press.

Koch, Christof. 2019b. "Consciousness in Cerebral Organoids—How Would We Know?" *University of California Television*. https://www.youtube.com/watch?v=vMYnzTn0G1k (accessed June 8, 2020).

Kohler, Robert E. 1972. "The Reception of Eduard Buchner's Discovery of Cell-Free Fermentation." *Journal of the History of Biology* 5:327–53.

Kolb, Vera M. 2019. *Handbook of Astrobiology*. Boca Raton, FL: CRC Press.

Kompanichenko, Vladimir N. 2019. "Exploring the Kamchatka Geothermal Region in the Context of Life's Beginning." *Life* 9. doi:10.3390/life9020041.

Kompanichenko, Vladimir N., Valery A. Poturay, and K. V. Shlufman. 2015. "Hydrothermal Systems of Kamchatka Are Models of the Prebiotic Environment." *Origins of Life and Evolution of Biospheres* 45:93–103.

Kopparapu, Ravi K., Eric T. Wolf, and Victoria S. Meadows. 2019. "Characterizing Exoplanet Habitability." arXiv:1911.04441.

Kothe, Erika. 1996. "Tetrapolar Fungal Mating Types: Sexes by the Thousands." *FEMS Microbiology Reviews* 18:65–87.

Lampert, Kathrin P., and M. Schartl. 2008. "The Origin and Evolution of a Unisexual Hybrid: *Poecilia formosa*." *Philosophical Transactions of the Royal Society B* 363:2901–9.

Lancaster, Madeline A., Magdalena Renner, Carol-Anne Martin, Daniel Wenzel, Louise S. Bicknell, Matthew E. Hurles, Tessa Homfray et al. 2013. "Cerebral Organoids Model Human Brain Development and Microcephaly." *Nature* 501:373–79.

Larsen, Gregory D. 2016. "The Peculiar Physiology of the Python." *Lab Animal* 45. doi:10.1038/laban.1027.

Laskowski, Kate L., Carolina Doran, David Bierbach, Jens Krause, and Max Wolf. 2019. "Naturally Clonal Vertebrates Are an Untapped Resource in Ecology and Evolution Research." *Nature Ecology & Evolution* 3:161–69.

Lasser, Karen E., Kristin Mickle, Sarah Emond, Rick Chapman, Daniel A. Ollendorf, and Steven D. Pearson. 2018. "Inotersen and Patisiran for Hereditary Transthyretin Amyloidosis: Effectiveness and Value." *Institute for Clinical and Economic Review*, October 4. https://icer-review.org/wp-content/uploads/2018/02/ICER_Amyloidosis_Final_Evidence_Report_100418.pdf (accessed June 8, 2020).

Lazcano, Antonio. 2016. "Alexandr I. Oparin and the Origin of Life: A Historical Reassessment of the Heterotrophic Theory." *Journal of Molecular Evolution* 83:214–22.

Lazcano, Antonio, and Jeffrey L. Bada. 2003. "The 1953 Stanley L. Miller Experiment: Fifty Years of Prebiotic Organic Chemistry." *Origins of Life and Evolution of the Biosphere* 33:235–42.

Lederberg, Joshua. 1967. "Science and Man . . . The Legal Start of Life." *Washington Post*, July 1, p. A13.

Lee, Patrick, and Robert P. George. 2001. "Embryology, Philosophy, & Human Dignity." *National Review*, August 9. https://web.archive.org/web/20011217063957/http://www.nationalreview.com/comment/comment-leeprint080901.html (accessed June 8, 2020).

Lenhoff, Howard M., and Sylvia G. Lenhoff. 1988. "Trembley's Polyps." *Scientific American* 258:108–13.

Lenhoff, Sylvia G., and Howard M. Lenhoff. 1986. *Hydra and the Birth of Experimental Biology—1744: Abraham Trembley's Memoires Concerning the Polyps*. Pacific Grove, CA: Boxwood Press.

Letelier, Juan-Carlos, María L. Cárdenas, and Athel Cornish-Bowden. 2011. "From *L'Homme Machine* to Metabolic Closure: Steps Towards Understanding Life." *Journal of Theoretical Biology* 286:100–113.

Levy, Steven. 1992. *Artificial Life: The Quest for a New Creation*. New York: Pantheon Books.

Lewis, Clive S. 1947. *The Abolition of Man: Or, Reflections on Education with Special Reference to the Teaching of English in the Upper Forms of School*. New York: Macmillan.

Lilley, Thomas M., Ian W. Wilson, Kenneth A. Field, DeeAnn M. Reeder, Megan E. Vodzak, Gregory G. Turner, Allen Kurta et al. 2020. "Genome-Wide Changes in Genetic Diversity in a Population of *Myotis lucifugus* Affected by White-Nose Syndrome." *G3* 10:2007–20.

Liu, Daniel. 2017. "The Cell and Protoplasm as Container, Object, and Substance, 1835–1861." *Journal of the History of Biology* 50:889–925.

Liu, Li, Jiajing Wang, Danny Rosenberg, Hao Zhao, György Lengyel, and Dani Nadel. 2018. "Fermented Beverage and Food Storage in 13,000 Y-Old Stone Mortars at Raqefet Cave, Israel: Investigating Natufian Ritual Feasting." *Journal of Archaeological Science: Reports* 21:783–93.

López-García, Purificación, and David Moreira. 2012. "Viruses in Biology." *Evolution: Education and Outreach* 5:389–98.

Luisi, Pier L. 1998. "About Various Definitions of Life." *Origins of Life and Evolution of the Biosphere* 28:613–22.

Lynn, Michael R. 2001. "Haller, Albrecht Von." *eLS*. doi:10.1038/npg.els.0002941.

Macdougall, Doug. 2019. *Endless Novelties of Extraordinary Interest: The Voyage of* H.M.S., *Challenger and the Birth of Modern Oceanography*. New Haven, CT: Yale University Press.

Machado, Calixto. 2005. "The First Organ Transplant from a Brain-Dead Donor." *Neurology* 64:1938–42.

Macleod, Gordon, Christopher McKeown, Allan J. Hall, and Michael J. Russell. 1994. "Hydrothermal and Oceanic pH Conditions of Possible Relevance to the Origin of Life." *Origins of Life and Evolution of the Biosphere* 24:19–41.

Maehle, Andreas-Holger. 1999. *Drugs on Trial: Experimental Pharmacology and Therapeutic Innovation in the Eighteenth Century*. Amsterdam: Rodopi.

Maienschein, Jane. 2014. "Politics in Your DNA." *Slate*, June 10. https://slate.com/technology/2014/06/personhood-movement-chimeras-how-biology-complicates-politics.html (accessed June 8, 2020).

Manninen, Bertha A. 2012. "Beyond Abortion: The Implications of Human Life Amendments." *Journal of Social Philosophy* 43:140–60.

Marchetto, Maria C. N., Cassiano Carromeu, Allan Acab, Diana Yu, Gene W. Yeo, Yangling Mu, Gong Chen et al. 2010. "A Model for Neural Development and Treatment of Rett Syndrome Using Human Induced Pluripotent Stem Cells." *Cell* 143:527–39.

Mariscal, Carlos, Ana Barahona, Nathanael Aubert-Kato, Arsev U. Aydinoglu, Stuart Bartlett, María L. Cárdenas, Kuhan Chandru et al. 2019. "Hidden Concepts in the History and Philosophy of Origins-of-Life Studies: A Workshop Report." *Origins of Life and Evolution of the Biosphere* 49:111–45.

Mariscal, Carlos, and W. F. Doolittle. 2018. "Life and Life Only: A Radical Alternative to Life Definitionism." *Synthese*. doi:10.1007/s11229-018-1852-2.

Marshall, Stuart M., Douglas Moore, Alastair R. G. Murray, Sara I. Walker, and Leroy Cronin. 2019. "Quantifying the Pathways to Life Using Assembly Spaces." arXiv:1907.04649.

Marshall, Stuart, et al. In preparation. "Identifying Molecules as Biosignatures with Assembly Theory and Mass Spectrometry." Manuscript.

Maruyama, Koscak. 1991. "The Discovery of Adenosine Triphosphate and the Establishment of Its Structure." *Journal of the History of Biology* 24:145–54.

McGrath, Larry. 2013. "Bergson Comes to America." *Journal of the History of Ideas* 74:599–620.

McGraw, Donald J. 1974. "Bye-Bye Bathybius: The Rise and Fall of a Marine Myth." *Bios* 45:164–71.

McInnis, Brian I. 2016. "Haller, Unzer, and Science as Process." In *The Early History of Embodied Cognition 1740–1920: The Lebenskraft-Debate and Radical Reality in German Science, Music, and Literature*. Edited by John A. McCarthy, Stephanie M. Hilger, Heather I. Sullivan, and Nicholas Saul. Leiden, Netherlands: Brill.

McKaughan, Daniel J. 2005. "The Influence of Niels Bohr on Max Delbrück: Revisiting the Hopes Inspired by 'Light and Life.'" *Isis* 96:507–29.

McKay, David S., Everett K. Gibson Jr., Kathie L. Thomas-Keprta, Hojatollah Vali, Christopher S. Romanek, Simon J. Clemett, Xavier D. F. Chillier et al.

1996. "Search for Past Life on Mars: Possible Relic Biogenic Activity in Martian Meteorite ALH84001." *Science* 273:924–30.

McNab, Brian K. 1969. "The Economics of Temperature Regulation in Neutropical Bats." *Comparative Biochemistry and Physiology* 31:227–68.

Meierhenrich, Uwe J. 2012. "Life in Its Uniqueness Remains Difficult to Define in Scientific Terms." *Journal of Biomolecular Structure and Dynamics* 29:641–42.

Mesci, Pinar, Angela Macia, Spencer M. Moore, Sergey A. Shiryaev, Antonella Pinto, Chun-Teng Huang, Leon Tejwani et al. 2018. "Blocking Zika Virus Vertical Transmission." *Scientific Reports* 8. doi:10.1038/s41598-018-19526-4.

Mesler, Bill, and H. J. Cleaves II. 2015. *A Brief History of Creation: Science and the Search for the Origin of Life*. New York: Norton.

Miller, Stanley L. 1974. "The First Laboratory Synthesis of Organic Compounds Under Primitive Earth Conditions." In *The Heritage Copernicus: Theories "Pleasing to the Mind."* Edited by Jerzy Neyman. Cambridge, MA: MIT Press.

Miller, Stanley L., J. W. Schopf, and Antonio Lazcano. 1997. "Oparin's 'Origin of Life': Sixty Years Later." *Journal of Molecular Evolution* 44:351–53.

Milshteyn, Daniel, Bruce Damer, Jeff Havig, and David Deamer. 2018. "Amphiphilic Compounds Assemble into Membranous Vesicles in Hydrothermal Hot Spring Water but Not in Seawater." *Life* 8. doi:10.3390/life8020011.

Miras, Haralampos N., Cole Mathis, Weimin Xuan, De-Liang Long, Robert Pow, and Leroy Cronin. 2019. "Spontaneous Formation of Autocatalytic Sets with Self-Replicating Inorganic Metal Oxide Clusters." *Proceedings of the National Academy of Sciences* 117:10699–705.

Mollaret, Pierre, and Maurice Goulon. 1959. "Le coma dépassé." *Revue Neurologique* 101:3–15.

Mommaerts, Wilfried F. 1992. "Who Discovered Actin?" *BioEssays* 14:57–59.

Moniruzzaman, Mohammad, Carolina A. Martinez-Gutierrez, Alaina R. Weinheimer, and Frank O. Aylward. 2020. "Dynamic Genome Evolution and Complex Virocell Metabolism of Globally-Distributed Giant Viruses." *Nature Communications* 11. doi:10.1038/s41467-020-15507-2.

Moore, Marianne S., Kenneth A. Field, Melissa J. Behr, Gregory G. Turner, Morgan E. Furze, Daniel W. F. Stern, Paul R. Allegra et al. 2018. "Energy Conserving Thermoregulatory Patterns and Lower Disease Severity in a Bat Resistant to the Impacts of White-Nose Syndrome." *Journal of Comparative Physiology B* 188:163–76.

Moradali, M. F., Shirin Ghods, and Bernd H. A. Rehm. 2017. "*Pseudomonas aeruginosa* Lifestyle: A Paradigm for Adaptation, Survival, and Persistence." *Frontiers in Cellular and Infection Microbiology* 7. doi:10.3389/fcimb.2017.00039.

Mortensen, Jens. 2020. "Six Months of Coronavirus: Here's Some of What We've Learned." *New York Times*, June 18. https://www.nytimes.com/article/coronavirus-facts-history.html (accessed July 25, 2020).

Moseley, Henry N. 1892. *Notes by a Naturalist: An Account of Observations Made During the Voyage of* H.M.S., *"Challenger" Round the World in the Years 1872–1876*. New York: Putnam.

Moss, Helen E., Lorraine K. Tyler, and Fábio Jennings. 1997. "When Leopards Lose Their Spots: Knowledge of Visual Properties in Category-Specific Deficits for Living Things." *Cognitive Neuropsychology* 14:901–50.

Moss, Ralph W. 1988. *Free Radical: Albert Szent-Gyorgyi and the Battle over Vitamin C.* New York: Paragon House.

"Mr. J. B. Butler Burke." *Times* (London), January 16, 1946, p. 6.

Mullen, Leslie. 2013. "Forming a Definition for Life: Interview with Gerald Joyce." Astrobiology Magazine, July 25. https://www.astrobio.net/origin-and-evolution-of-life/forming-a-definition-for-life-interview-with-gerald-joyce/ (accessed June 8, 2020).

Muller, Hermann J. 1929. "The Gene as the Basis of Life." *Proceedings of the International Congress of Plant Sciences* 1:879–921.

Murray, John. 1876. "Preliminary Reports to Professor Wyville Thomson, F.R.S., Director of the Civilian Scientific Staff, on Work Done on Board the 'Challenger.'" *Proceedings of the Royal Society* 24:471–544.

Nair-Collins, Michael. 2018. "A Biological Theory of Death: Characterization, Justification, and Implications." *Diametros* 55:27–43.

Nair-Collins, Michael, Jesse Northrup, and James Olcese. 2016. "Hypothalamic-Pituitary Function in Brain Death: A Review." *Journal of Intensive Care Medicine* 31:41–50.

Nairne, James S., Joshua E. VanArsdall, and Mindi Cogdill. 2017. "Remembering the Living: Episodic Memory Is Tuned to Animacy." *Current Directions in Psychological Science* 26:22–27.

Navarro-Costa, Paulo, and Rui G. Martinho. 2020. "The Emerging Role of Transcriptional Regulation in the Oocyte-to-Zygote Transition." *PLoS Genetics* 16. doi:10.1371/journal.pgen.1008602.

Neaves, William. 2017. "The Status of the Human Embryo in Various Religions." *Development* 144:2541–43.

Needham, Joseph. 1925. "The Philosophical Basis of Biochemistry." *Monist* 35:27–48.

Nicholson, Daniel J., and Richard Gawne. 2015. "Neither Logical Empiricism Nor Vitalism, but Organicism: What the Philosophy of Biology Was." *History and Philosophy of the Life Sciences* 37:345–81.

Nobis, Nathan, and Kristina Grob. 2019. *Thinking Critically About Abortion: Why Most Abortions Aren't Wrong & Why All Abortions Should Be Legal.* Open Philosophy Press.

Noonan, John T., Jr. 1967. "Abortion and the Catholic Church: A Summary History." *American Journal of Jurisprudence* 12:85–131.

Normandin, Sebastian, and Charles T. Wolfe. 2013. *Vitalism and the Scientific Image in Post-Enlightenment Life Science, 1800–2010.* New York: Springer.

Oberhaus, Daniel. 2019. "A Crashed Israeli Lunar Lander Spilled Tardigrades on the Moon." *Wired*, August 5. https://www.wired.com/story/a-crashed-israeli-lunar-lander-spilled-tardigrades-on-the-moon/ (accessed June 8, 2020).

"Obituary Notices of Fellows Deceased." *Proceedings of the Royal Society*, January 1, 1895. doi:10.1098/rspl.1895.0002.

O'Brien, Charles F. 1970. "*Eozoön canadense*: 'The Dawn Animal of Canada,'" *Isis* 61:206–23.

Oettmeier, Christina, Klaudia Brix, and Hans-Günther Döbereiner. 2017. "*Physarum polycephalum*—A New Take on a Classic Model System." *Journal of Physics D* 50. doi:10.1088/1361-6463/aa8699.

Ohl, Christiane, and Wilhelm Stockem. 1995. "Distribution and Function of Myosin II as a Main Constituent of the Microfilament System in *Physarum polycephalum*." *European Journal of Protistology* 31:208–22.

Olby, Robert. 2009. *Francis Crick: Hunter of Life's Secrets*. Cold Spring Harbor, NY: Cold Spring Harbor Laboratory Press.

Oparin, Alexander I. 1924. "The Origin of Life." In *The Origin of Life*. Edited by John D. Bernal. Cleveland, OH: World Publishing Company.

Oparin, Alexander I. 1938. *The Origin of Life*. New York: Macmillan.

Orgel, Leslie E. 1968. "Evolution of the Genetic Apparatus." *Journal of Molecular Biology* 38:381–93.

"Oriental Memoirs." 1814. *Monthly Magazine* 36:577–618.

"Origin of Life, The." *Cambridge Independent Press*, June 23, 1905, p. 3.

Packard, Alpheus S. 1876. *Life Histories of Animals, Including Man: Or, Outlines of Comparative Embryology*. New York: Henry Holt.

Palacios, Ensor R., Adeel Razi, Thomas Parr, Michael Kirchhoff, and Karl Friston. 2020. "On Markov Blankets and Hierarchical Self-Organisation." *Journal of Theoretical Biology* 486:110089.

Paleos, Constantinos M. 2015. "A Decisive Step Toward the Origin of Life." *Trends in Biochemical Sciences* 40:487–88.

Parrilla-Gutierrez, Juan M., Soichiro Tsuda, Jonathan Grizou, James Taylor, Alon Henson, and Leroy Cronin. 2017. "Adaptive Artificial Evolution of Droplet Protocells in a 3D-Printed Fluidic Chemorobotic Platform with Configurable Environments." *Nature Communications* 8. doi:10.1038/s41467-017-01161-8.

Peabody, C. A. 1882. "Marriage and Its Duties." *Daily Journal* (Montpelier, VT), November 8, p. 4.

Peattie, Donald C. 1950. *A Natural History of Trees of Eastern and Central North America*. Boston: Houghton Mifflin.

Peck, Carol J., and Nels R. Lersten. 1991a. "Samara Development of Black Maple (*Acer saccharum* Ssp. *nigrum*) with Emphasis on the Wing." *Canadian Journal of Botany* 69:1349–60.

Peck, Carol J., and Nels R. Lersten. 1991b. "Gynoecial Ontogeny and Morphology, and Pollen Tube Pathway in Black Maple, *Acer saccharum* Ssp. *nigrum* (*Aceraceae*)." *American Journal of Botany* 78:247–59.

Penning, David A., Schuyler F. Dartez, and Brad R. Moon. 2015. "The Big Squeeze: Scaling of Constriction Pressure in Two of the World's Largest Snakes, *Python reticulatus* and *Python molurus bivittatus*." *Journal of Experimental Biology* 218:3364–67.

Peretó, Juli, Jeffrey L. Bada, and Antonio Lazcano. 2009. "Charles Darwin and the Origin of Life." *Origins of Life and Evolution of Biospheres* 39:395–406.

Perry, Blair W., Audra L. Andrew, Abu H. M. Kamal, Daren C. Card, Drew R. Schield, Giulia I. M. Pasquesi, Mark W. Pellegrino et al. 2019. "Multi-Species Comparisons of Snakes Identify Coordinated Signalling Networks Underlying Post-Feeding Intestinal Regeneration." *Proceedings of the Royal Society B* 286. doi:10.1098/rspb.2019.0910.

Peters, Philip G., Jr. 2006. "The Ambiguous Meaning of Human Conception." *UC Davis Law Review* 40:199–228.

Pettitt, Paul B. 2018. "Hominin Evolutionary Thanatology from the Mortuary to Funerary Realm: The Palaeoanthropological Bridge Between Chemistry and Culture." *Philosophical Transactions of the Royal Society B* 373. doi:10.1098/rstb.2018.0212.

Pfeiffer, Burkard, and Frieder Mayer. 2013. "Spermatogenesis, Sperm Storage and Reproductive Timing in Bats." *Journal of Zoology* 289:77–85.

Phillips, R. 2020. "Schrodinger's 'What is Life?' at 75: Back to the Future." Manuscript.

Pierpont, W. S. 1999. "Norman Wingate Pirie: 1 July 1907–29 March 1997." *Biographical Memoirs of Fellows of the Royal Society* 45:399–415.

Pirie, Norman W. 1937. "The Meaninglessness of the Terms Life and Living." In *Perspectives in Biochemistry: Thirty-One Essays Presented to Sir Frederick Gowland Hopkins by Past and Present Members of His Laboratory*. Edited by Joseph Needham and David E. Green. Cambridge: Cambridge University Press.

Points, Laurie J., James W. Taylor, Jonathan Grizou, Kevin Donkers, and Leroy Cronin. 2018. "Artificial Intelligence Exploration of Unstable Protocells Leads to Predictable Properties and Discovery of Collective Behavior." *Proceedings of the National Academy of Sciences* 115. doi:10.1073/pnas.1711089115.

Poltak, Steffen R., and Vaughn S. Cooper. 2011. "Ecological Succession in Long-Term Experimentally Evolved Biofilms Produces Synergistic Communities." *ISME Journal* 5:369–78.

Popa, Radu. 2004. *Between Necessity and Probability: Searching for the Definition and Origin of Life*. Berlin: Springer-Verlag.

Porcar, Manuel, and Juli Peretó. 2018. "Creating Life and the Media: Translations and Echoes." *Life Sciences, Society and Policy* 14. doi:10.1186/s40504-018-0087-9.

Postberg, Frank, Nozair Khawaja, Bernd Abel, Gael Choblet, Christopher R. Glein, Murthy S. Gudipati, Bryana L. Henderson et al. 2018. "Macromolecular Organic Compounds from the Depths of Enceladus." *Nature* 558:564–68.

Pratama, Akbar A., and Jan D. Van Elsas. 2018. "The 'Neglected' Soil Virome—Potential Role and Impact." *Trends in Microbiology* 26:649–62.

President's Commission for the Study of Ethical Problems in Medicine and Biomedical and Behavioral Research. 1981. *Defining Death: A Report on the Medical, Legal and Ethical Issues in the Determination of Death*. Washington, D.C.: U.S. Government Printing Office.

Quick, Joshua, Nicholas J. Loman, Sophie Duraffour, Jared T. Simpson, Ettore Severi, Lauren Cowley, Joseph A. Bore et al. 2016. "Real-Time, Portable Genome Sequencing for Ebola Surveillance." *Nature* 530:228–32.

Rajamani, Sudha, Alexander Vlassov, Seico Benner, Amy Coombs, Felix Olasagasti, and David Deamer. 2008. "Lipid-Assisted Synthesis of RNA-Like Polymers from Mononucleotides." *Origins of Life and Evolution of Biospheres* 38:57–74.

Rall, Jack A. 2018. "Generation of Life in a Test Tube: Albert Szent-Gyorgyi, Bruno Straub, and the Discovery of Actin." *Advances in Physiology Education* 42:277–88.

Ramberg, Peter J. 2000. "The Death of Vitalism and the Birth of Organic Chemistry: Wohler's Urea Synthesis and the Disciplinary Identity of Organic Chemistry." *Ambix*, 47170–95

Rankin, Mark. 2013. "Can One Be Two? A Synopsis of the Twinning and Personhood Debate." *Monash Bioethics Review* 31:37–59.

Ratcliff, Marc J. 2004. "Abraham Trembley's Strategy of Generosity and the Scope of Celebrity in the Mid-Eighteenth Century." *Isis* 95:555–75.

Ray, Subash K., Gabriele Valentini, Purva Shah, Abid Haque, Chris R. Reid, Gregory F. Weber, and Simon Garnier. 2019. "Information Transfer During Food Choice in the Slime Mold *Physarum polycephalum*." *Frontiers in Ecology and Evolution* 7:1–11.

Reed, Charles B. 1915. *Albrecht Von Haller: A Physician—Not Without Honor*. Chicago: Chicago Literary Club.

Rehbock, Philip F. 1975. "Huxley, Haeckel, and the Oceanographers: The Case of *Bathybius haeckelii*." *Isis* 66:504–33.

Reid, Chris R., Hannelore MacDonald, Richard P. Mann, James A. R. Marshall, Tanya Latty, and Simon Garnier. 2016. "Decision-Making Without a Brain: How an Amoeboid Organism Solves the Two-Armed Bandit." *Journal of the Royal Society Interface* 13. doi:10.1098/rsif.2016.0030.

Reid, Chris R., Simon Garnier, Madeleine Beekman, and Tanya Latty. 2015. "Information Integration and Multiattribute Decision Making in Non-Neuronal Organisms." *Animal Behaviour* 100:44–50.

Reid, Chris R., Tanya Latty, Andrey Dussutour, and Madeleine Beekman. 2012. "Slime Mold Uses an Externalized Spatial 'Memory' to Navigate in Complex Environments." *Proceedings of the National Academy of Sciences* 109:17490–94.

Reinhold, Robert. 1968. "Harvard Panel Asks Definition of Death Be Based on Brain." *New York Times*, August 5. https://www.nytimes.com/1968/08/05/archives/harvard-panel-asks-definition-of-death-be-based-on-brain-death.html (accessed June 8, 2020).

Rice, Amy L. 1983. "Thomas Henry Huxley and the Strange Case Of *Bathybius haeckelii*: A Possible Alternative Explanation." *Archives of Natural History* 11:169–80.

Robinson, Denis M. 1988. "Reminiscences on Albert Szent-Györgyi." *Biological Bulletin* 174:214–33.

Rochlin, Kate, Shannon Yu, Sudipto Roy, and Mary K. Baylies. 2010. "Myoblast Fusion: When It Takes More to Make One." *Developmental Biology* 341: 66–83.

Roe, Shirley A. 1981. *Matter, Life, and Generation: 18th-Century Embryology and the Haller-Wolff Debate*. Cambridge, UK: Cambridge University Press.

Rogalla, Horts, and Peter H. Kes, editors. *100 Years of Superconductivity.* London: Taylor & Francis.

Roger, Jacques. 1997. *Buffon: A Life in Natural History*. Translated by Sarah L. Bonnefoi. Ithaca, NY: Cornell University Press.

Rosa-Salva, Orsola, Uwe Mayer, and Giorgio Vallortigara. 2015. “Roots of a Social Brain: Developmental Models of Emerging Animacy-Detection Mechanisms.” *Neuroscience & Biobehavioral Reviews* 50:150–68.

Rößler, Hole. 2013. “Character Masks of Scholarship: Self-Representation and Self-Experiment as Practices of Knowledge Around 1770.” In *Scholars in Action: The Practice of Knowledge and the Figure of the Savant in the 18th Century*. Edited by André Holenstein, Hubert Steinke, and Martin Stuber. Leiden, Netherlands: Brill.

Ruggiero, Angela. 2018. “Jahi McMath: Funeral Honors Young Teen Whose Brain Death Captured World’s Attention.” *Mercury News* (San Jose, CA), July 6. https://www.mercurynews.com/2018/07/06/jahi-mcmath-funeral-honors-young-teen-whose-brain-death-captured-worlds-attention/ (accessed June 8, 2020).

Rummel, Andrea D., Sharon M. Swartz, and Richard L. Marsh. 2019. “Warm Bodies, Cool Wings: Regional Heterothermy in Flying Bats.” *Biology Letters* 15. doi:10.1098/rsbl.2019.0530.

Rupke, Nicolaas A. 1976. “*Bathybius haeckelii* and the Psychology of Scientific Discovery: Theory Instead of Observed Data Controlled the Late 19th Century ‘Discovery’ of a Primitive Form of Life.” *Studies in History and Philosophy of Science Part A* 7:53–62.

Russell, Michael J. 2019. “Prospecting for Life.” *Interface Focus* 9. doi:10.1098/rsfs.2019.0050.

Rutz, Christian, Matthias-Claudio Loretto, Amanda E. Bates, Sarah C. Davidson, Carlos M. Duarte, Walter Jetz, Mark Johnson et al. 2020. “COVID-19 Lockdown Allows Researchers to Quantify the Effects of Human Activity on Wildlife.” *Nature Ecology & Evolution*. doi:10.1038/s41559-020-1237-z.

Sagan, Carl, and Joshua Lederberg. 1976. “The Prospects for Life on Mars: A Pre-Viking Assessment.” *Icarus* 28:291–300.

Samartzidou, Hrissi, Mahsa Mehrazin, Zhaohui Xu, Michael J. Benedik, and Anne H. Delcour. 2003. “Cadaverine Inhibition of Porin Plays a Role in Cell Survival at Acidic pH.” *Journal of Bacteriology* 185:13–19.

Satterly, John. 1939. “The Postprandial Proceedings of the Cavendish Society I.” *American Journal of Physics* 7:179–85.

Schlenk, Fritz. 1987. “The Ancestry, Birth and Adolescence of Adenosine Triphosphate.” *Trends in Biochemical Sciences* 12:367–68.

Schmalian, Jörg. 2010. “Failed Theories of Superconductivity.” *Modern Physics Letters B* 24:2679–91.

Scholl, Brian J., and Patrice D. Tremoulet. 2000. “Perceptual Causality and Animacy.” *Trends in Cognitive Sciences* 4:299–309.

Schrödinger, Erwin. 2012. *What Is Life?* Cambridge: Cambridge University Press.

Secor, Stephen M., and Jared Diamond. 1995. "Adaptive Responses to Feeding in Burmese Pythons: Pay Before Pumping." *Journal of Experimental Biology* 198:1313–25.

Secor, Stephen M., and Jared Diamond. 1998. "A Vertebrate Model of Extreme Physiological Regulation." *Nature* 395:659–62.

Secor, Stephen M., Eric D. Stein, and Jared Diamond. 1994. "Rapid Upregulation of Snake Intestine in Response to Feeding: A New Model of Intestinal Adaptation." *American Journal of Physiology* 266:G695–705.

Setia, Harpreet, and Alysson R. Muotri. 2019. "Brain Organoids as a Model System for Human Neurodevelopment and Disease." *Seminars in Cell and Developmental Biology* 95:93–97.

Setten, Ryan L., John J. Rossi, and Si-Ping Han. 2019. "The Current State and Future Directions of RNAi-Based Therapeutics." *Nature Reviews Drug Discovery* 18:421–46.

Shahar, Anat, Peter Driscoll, Alycia Weinberger, and George Cody. 2019. "What Makes a Planet Habitable?" *Science* 364:434–35.

Shewmon, D. A. 2018. "The Case of Jahi McMath: A Neurologist's View." *Hastings Center Report* 48:S74–76.

Simkulet, William. 2017. "Cursed Lamp: The Problem of Spontaneous Abortion." *Journal of Medical Ethics*. doi:10.1136/medethics-2016-104018.

Simpson, Bob. 2018. "Death." *Cambridge Encyclopedia of Anthropology*, July 23. http://doi.org/10.29164/18death (accessed June 8, 2020).

Sloan, Philip R., and Brandon Fogel. 2011. *Creating a Physical Biology: The Three-Man Paper and Early Molecular Biology*. Chicago: University of Chicago Press.

Slowik, Edward. 2017. "Descartes' Physics." *Stanford Encyclopedia of Philosophy*, August 22. https://plato.stanford.edu/archives/fall2017/entries/descartes-physics/ (accessed July 25, 2020).

Slutsky, Arthur S. 2015. "History of Mechanical Ventilation: From Vesalius to Ventilator-Induced Lung Injury." *American Journal of Respiratory and Critical Care Medicine* 191:1106–15.

Smith, Kelly C. 2018. "Life as Adaptive Capacity: Bringing New Life to an Old Debate." *Biological Theory* 13:76–92.

Srivathsan, Amrita, Emily Hartop, Jayanthi Puniamoorthy, Wan T. Lee, Sujatha N. Kutty, Olavi Kurina, and Rudolf Meier. 2019. "Rapid, Large-Scale Species Discovery in Hyperdiverse Taxa Using 1D MinION Sequencing." *BMC Biology* 17. doi:10.1186/s12915-019-0706-9.

Steigerwald, Joan. 2019. *Experimenting at the Boundaries of Life: Organic Vitality in Germany Around 1800*. Pittsburgh, PA: University of Pittsburgh Press.

Steinke, Hubert. 2005. *Irritating Experiments: Haller's Concept and the European Controversy on Irritability and Sensibility, 1750–90*. Amsterdam: Rodopi.

Stephenson, Andrew G. 1981. "Flower and Fruit Abortion: Proximate Causes and Ultimate Functions." *Annual Review of Ecology and Systematics* 12:253–79.

Stiles, Joan, and Terry L. Jernigan. 2010. "The Basics of Brain Development." *Neuropsychology Review* 20:327–48.

Stott, Rebecca. 2012. *Darwin's Ghosts: The Secret History of Evolution*. New York: Spiegel & Grau.

Strauss, Bernard S. 2017. "A Physicist's Quest in Biology: Max Delbrück and 'Complementarity.'" *Genetics* 206:641–50.

Strick, James E. 2009. "Darwin and the Origin of Life: Public Versus Private Science." *Endeavour* 33:148–51.

Subramanian, Samanth. 2020. *A Dominant Character: The Radical Science and Restless Politics of J. B. S. Haldane*. New York: Norton.

Sullivan, Janet R. 1983. "Comparative Reproductive Biology of *Acer pensylvanicum* and *A. spicatum* (*Aceraceae*)." *American Journal of Botany* 70:916–24.

Surman, Andrew J., Marc R. Garcia, Yousef M. Abul-Haija, Geoffrey J. T. Cooper, Piotr S. Gromski, Rebecca Turk-MacLeod, Margaret Mullin et al. 2019. "Environmental Control Programs the Emergence of Distinct Functional Ensembles from Unconstrained Chemical Reactions." *Proceedings of the National Academy of Sciences* 116. doi:10.1073/pnas.1813987116.

Sutton, Geoffrey. 1984. "The Physical and Chemical Path to Vitalism: Xavier Bichat's Physiological Researches on Life and Death." *Bulletin of the History of Medicine* 58:53–71.

Swartz, Mimi. 1996. "It Came from Outer Space." *Texas Monthly*, November. https://www.texasmonthly.com/articles/it-came-from-outer-space/ (accessed July 25, 2020).

Sweet, William H. 1978. "Brain Death." *New England Journal of Medicine* 299:410–22.

Symonds, Neville. 1988. "Schrödinger and Delbrück: Their Status in Biology." *Trends in Biochemical Sciences* 13:232–34.

Szabo, Liz. 2014. "Ethicists Criticize Treatment of Teen, Texas Patient." *USA Today*, January 9. https://www.usatoday.com/story/news/nation/2014/01/09/ethicists-criticize-treatment-brain-dead-patients/4394173/ (accessed June 8, 2020).

Szent-Györgyi, Albert. 1948. *Nature of Life: A Study on Muscle*. New York: Academic Press.

Szent-Györgyi, Albert. 1963. "Lost in the Twentieth Century." *Annual Review of Biochemistry* 32:1–14.

Szent-Györgyi, Albert. 1972. "What Is Life?" In *Biology Today*. Edited by John H. Painter, Jr. Del Mar, CA: CRM Books.

Szent-Györgyi, Albert. 1977. "The Living State and Cancer." *Proceedings of the National Academy of Sciences* 74:2844–47.

Tamura, Koji. 2016. "The Genetic Code: Francis Crick's Legacy and Beyond." *Life* 6:36.

Taubner, Ruth-Sophie, Patricia Pappenreiter, Jennifer Zwicker, Daniel Smrzka, Christian Pruckner, Philipp Kolar, Sébastien Bernacchi et al. 2018. "Biological Methane Production Under Putative Enceladus-Like Conditions." *Nature Communications* 9:748.

Taylor, William R. 1920. *A Morphological and Cytological Study of Reproduction in the Genus Acer*. Philadelphia: University of Pennsylvania.

Thomson, Charles W. 1869. "XIII. On the Depths of the Sea." *Annals and Magazine of Natural History* 4:112–24.

Thomson, Joseph J. 1906. "Some Applications of the Theory of Electric Discharge Through Gases to Spectroscopy." *Nature* 73:495–99.

Tirard, Stéphane. 2017. "J. B. S. Haldane and the Origin of Life." *Journal of Genetics* 96:735–39.

Trifonov, Edward N. 2011. "Vocabulary of Definitions of Life Suggests a Definition." *Journal of Biomolecular Structure and Dynamics* 29:259–66.

Trujillo, Cleber A., Richard Gao, Priscilla D. Negraes, Jing Gu, Justin Buchanan, Sebastian Preissl, Allen Wang et al. 2019. "Complex Oscillatory Waves Emerging from Cortical Organoids Model Early Human Brain Network Development." *Cell Stem Cell* 25:558–69.e7.

Truog, Robert D. 2018. "Lessons from the Case of Jahi McMath." *Hastings Center Report* 48:S70–73.

Vallortigara, Giorgio, and Lucia Regolin. 2006. "Gravity Bias in the Interpretation of Biological Motion by Inexperienced Chicks." *Current Biology* 16:R279–80.

Van Kranendonk, Martin J., David W. Deamer, and Tara Djokic. 2017. "Life Springs." *Scientific American* 317:28–35.

Van Lawick-Goodall, Jane. 1968. "The Behaviour of Free-Living Chimpanzees in the Gombe Stream Reserve." *Animal Behaviour Monographs* 1:161–311.

Van Lawick-Goodall, Jane. 1971. *In the Shadow of Man*. Boston: Houghton Mifflin.

Vartanian, Aram. 1950. "Trembley's Polyp, La Mettrie, and Eighteenth-Century French Materialism." *Journal of the History of Ideas* 11:259–86.

Vastenhouw, Nadine L., Wen X. Cao, and Howard D. Lipshitz. 2019. "The Maternal-to-Zygotic Transition Revisited." *Development* 146. doi:10.1242/dev.161471.

Vázquez-Diez, Cayetana, and Greg FitzHarris. 2018. "Causes and Consequences of Chromosome Segregation Error in Preimplantation Embryos." *Reproduction* 155:R63–76.

"Viking I Lands on Mars." *ABC News*, July 20, 1976. https://www.youtube.com/watch?v=gZjCfNvx9m8 (accessed June 8, 2020).

Villavicencio, Raphael T. 1998. "The History of Blue Pus." *Journal of the American College of Surgeons* 187:212–16.

Vitas, Marko, and Andrej Dobovišek. 2019. "Towards a General Definition of Life." *Origins of Life and Evolution of Biospheres* 49:77–88.

Vitturi, Bruno K., and Wilson L. Sanvito. 2019. "Pierre Mollaret (1898–1987)." *Journal of Neurology* 266:1290–91.

Voigt, Christian C., Winifred F. Frick, Marc W. Holderied, Richard Holland, Gerald Kerth, Marco A. R. Mello, Raina K. Plowright et al. 2017. "Principles and Patterns of Bat Movements: From Aerodynamics to Ecology." *Quarterly Review of Biology* 92:267–87.

Waddington, Conrad H. 1967. "No Vitalism for Crick." *Nature* 216:202–3.

Wakefield, Priscilla. 1816. *Instinct Displayed, in a Collection of Well-Authenticated Facts, Exemplifying the Extraordinary Sagacity of Various Species of the Animal Creation*. Boston: Flagg & Gould.

Walker, Sara I. 2017. "Origins of Life: A Problem for Physics, a Key Issues Review." *Reports on Progress in Physics* 80. doi:10.1088/1361-6633/aa7804.

Walker, Sara I. 2018. "Bio from Bit." In *Wandering Towards a Goal: How Can Mindless Mathematical Laws Give Rise to Aims and Intention?* Edited by Anthony Aguirre, Brendan Foster, and Zeeya Merali. Cham, Switzerland: Springer International Publishing.

Walker, Sara I., and Paul C. W. Davies. 2012. "The Algorithmic Origins of Life." *Journal of the Royal Society Interface* 10. doi:10.1098/rsif.2012.0869.

Walker, Sara I., Hyunju Kim, and Paul C. W. Davies. 2016. "The Informational Architecture of the Cell." *Philosophical Transactions of the Royal Society A* 374. doi:10.1098/rsta.2015.0057.

Walker, Sara I., William Bains, Leroy Cronin, Shiladitya DasSarma, Sebastian Danielache, Shawn Domagal-Goldman, Betul Kacar et al. 2018. "Exoplanet Biosignatures: Future Directions." *Astrobiology* 18:779–824.

Webb, Richard L., and Paul A. Nicoll. 1954. "The Bat Wing as a Subject for Studies in Homeostasis of Capillary Beds." *Anatomical Record* 120:253–63.

Welch, G. R. 1995. "T. H. Huxley and the 'Protoplasmic Theory of Life': 100 Years Later." *Trends in Biochemical Sciences* 20:481–85.

Westall, Frances, and André Brack. 2018. "The Importance of Water for Life." *Space Science Reviews* 214. doi:10.1007/s11214-018-0476-7.

Wijdicks, Eelco F. M. 2003. "The Neurologist and Harvard Criteria for Brain Death." *Neurology* 61:970–76.

Willis, Craig K. R. 2017. "Trade-offs Influencing the Physiological Ecology of Hibernation in Temperate-Zone Bats." *Integrative and Comparative Biology* 57:1214–24.

Wilson, Edmund B. 1923. *The Physical Basis of Life*. New Haven, CT: Yale University Press.

WSFA Staff. 2019. "Rape, Incest Exceptions Added to Abortion Bill." *WBRC FOX6 News*, May 8. https://www.wbrc.com/2019/05/08/rape-incest-exceptions-added-abortion-bill/ (accessed July 25, 2020).

Xavier, Joana C., Wim Hordijk, Stuart Kauffman, Mike Steel, and William F. Martin. 2020. "Autocatalytic Chemical Networks at the Origin of Metabolism." *Proceedings of the Royal Society B* 287. doi:10.1098/rspb.2019.2377.

Yashina, Svetlana, Stanislav Gubin, Stanislav Maksimovich, Alexandra Yashina, Edith Gakhova, and David Gilichinsky. 2012. "Regeneration of Whole Fertile Plants from 30,000-Y-Old Fruit Tissue Buried in Siberian Permafrost." *Proceedings of the National Academy of Sciences* 109:4008–13.

Yoxen, Edward J. 1979. "Where Does Schroedinger's 'What is Life?' Belong in the History of Molecular Biology?" *History of Science* 17:17–52.

Zammito, John H. 2018. *The Gestation of German Biology: Philosophy and Physiology from Stahl to Schelling*. Chicago: University of Chicago Press.

Zimmer, Carl. 1995. "First Cell." *Discover*, October 31. https://www.discovermagazine.com/the-sciences/first-cell (accessed June 8, 2020).

Zimmer, Carl. 2007. "The Meaning of Life." *Seed*, September 4. https://carlzimmer.com/the-meaning-of-life-437/ (accessed July 25, 2020).

Zimmer, Carl. 2011. "Darwin Under the Microscope: Witnessing Evolution in Microbes." In *In the Light of Evolution: Essays from the Laboratory and Field*. Edited by Jonathan B. Losos. New York: Macmillan.

Zimmer, Carl. 2021. *A Planet of Viruses*. Third edition. Chicago: University of Chicago Press.

ACKNOWLEDGMENTS

The seed of this book was a conversation I had with Ben Lillie, who owns the venue Caveat in Manhattan. On a walk in the Lower East Side, I suggested that he host a series of talks about life. It would be an easy thing to put together, I assured him, and he wondered if I would try. It turned out to be harder than I imagined, but it was certainly worth the effort. I got to speak with eight deep thinkers about life: Sara Walker, Carlos Mariscal, Jim Cleaves, Caleb Scharf, Jeremy England, Steven Benner, Donato Giovannelli, and Kate Adamala. Ben and I turned the conversations into a podcast (carlzimmer.com/podcasts), supported by Science Sandbox, a Simons Foundation initiative. Yet that experience did not satisfy my curiosity. It only deepened it. When I mentioned the idea of a book about life to my colleague Ed Yong, he said that was a book he would want to read. So I'd like to thank everyone who helped to guide me to the starting line of this marathon.

The Sloan Foundation kindly provided a fellowship that supported this book during uncertain times. Michael Mason and Celia Dugger at the *New York Times* enabled me to do preliminary reporting. My research for each chapter was made possible through the generosity of others. For the introduction, I'm grateful to Luis Campos for discussions about John Butler Burke. Allyson Muotri, Cleber Trujillo, Priscilla Negraes, and their colleagues introduced me to the mysteries of organoids. Jeantine Lunshof and I. Glenn Cohen helped me think about the ethics of the beginnings of life, while Karen Schindler guided me through oocyte development.

I'd like to thank Stephen Secor and David Nelson for the chance to commune with snakes, and Simon Garnier and his students for growing slime molds for me. For the chance to see bats in winter, I'd like to thank Carl Herzog, Katelyn Ritzko, and Lori Severino of the New York Department of Environmental Conservation, as well as Alexander Novick of the Lake George Land Conservancy. Thanks also to Sharon Schwartz for letting me visit her lab at Brown and talk about bats in flight. Rachel Spicer taught me about trees, and Isabel Ott, Abigail Matela, Vaughn Cooper, and Paul Turner gave me the chance to learn about evolution the hard way.

For the history of biology, I'm grateful to Patrick Anthony for helping me learn about Albrecht von Haller, and to Gary Wnek for putting me onto Albert Szent-Györgyi. I'd like to thank David Deamer for many years of conversation, and Laurie Barge for showing me how to make chemical gardens. Thanks to Lee Cronin for giving me a glimpse into the world of robot chemistry even as a pandemic left us both stuck at home.

Kate Adamala, Luis Campos, and Rob Phillips all kindly read over the entire manuscript. Thanks also go to my intrepid fact-checkers, Lorenzo Arvanitis, Britt Bistis, Nakeirah Christie, Kelly Farley, Lori Jia, Matt Kristoffersen, Anin Luo, and Krish Maypole.

I'm grateful to Stephen Morrow, my editor at Dutton, for seeing the shape of something new even when it was hard to limn, as well as my agent, Eric Simonoff, whose instinct for the right project remains as diamond-edged as ever.

Finally, my deepest thanks go to my family—to my daughters, Charlotte and Veronica, for handling a tough pandemic year with aplomb, and most of all to my wife, Grace. I often wonder what good fortune allowed me to spend my life with her. My love for her outstrips the words I can use to express it, so that every draft of acknowledgment begs for a rewrite.

INDEX

PROLOGUE

The worst scares of my life have usually come in unfamiliar places. I still panic a bit when I remember traveling into a Sumatran jungle, only to discover my brother, Ben, had dengue fever. I lose a bit of breath any time I think about a night in Bujumbura when a friend and I got mugged. My fingers still curl when I recall a fossil-mad paleontologist leading me to the slick mossy edge of a Newfoundland cliff in search of Precambrian life. But the greatest scare of all, the one that made the world suddenly unfamiliar, swept over me while I was sitting with my wife, Grace, in the comfort of an obstetrician's office.

Grace was pregnant with our first child, and our obstetrician had insisted we meet with a genetics counselor. We didn't see the point. We felt untroubled in being carried along into the future, wherever we might end up. We knew Grace had a second heartbeat inside her, a healthy one, and that seemed enough to know. We didn't even want to find out if the baby was a girl or a boy. We would just debate names in two columns: Liam or Henry, Charlotte or Catherine.

Still, our doctor insisted. And so one afternoon we went to an office in lower Manhattan, where we sat down with a middle-aged woman, perhaps a decade older than us. She was cheerful and clear, talking about our child's health beyond what the thrum of a heartbeat could tell us. We were politely cool, wanting to end this appointment as soon as possible.

We had already talked about the risks we faced starting a family in our thirties, the climbing odds that our children might have Down syndrome. We agreed that we'd deal with whatever challenges our child faced. I felt proud of my commitment. But now, when I look back at my younger self, I'm not so impressed. I didn't know anything at the time about what it's actually like raising a child with Down syndrome. A few years later, I would get to know some parents who were doing just that. Through them, I would get a glimpse of that life: of round after round of heart surgeries, of the struggle to teach children how to behave with outsiders, of the worries about a child's future after one's own death.

But as we sat that day with our genetics counselor, I was still blithe, still confident. The counselor could tell we didn't want to be there, but she managed to keep the conversation alive. Down syndrome was not the only thing expectant parents should think about, she said. It was possible that the two of us carried genetic variations that we could pass down to our child, causing other disorders. The counselor took out a piece of paper and drew a family tree, to show us how genes were inherited.

"You don't have to explain all that to us," I assured her. After all, I wrote about things like genes for a living. I didn't need a high school lecture.

"Well, let me ask you a little about your family," she replied.

It was 2001. A few months beforehand, two geneticists had come to the White House to stand next to President Bill Clinton for an announcement. "We are here to celebrate the completion of the first survey of the entire human genome," Clinton said. "Without a doubt, this is the most important, most wondrous map ever produced by humankind."

The "entire human genome" that Clinton was hailing didn't come from any single person on Earth. It was an error-ridden draft, a collage of genetic material pieced together from a mix of people. And it had cost $3 billion. Rough as it was, its completion did mark a historical moment in the history of science. A rough map is far better than no map at all. Scientists began to compare the human genome to the genomes of other species, in order to learn on a molecular level how we evolved from common ancestors. They could examine the twenty thousand–odd genes that encode human proteins, one at a time, to learn about how they helped make a human and the part they played in how we get sick.

In 2001, Grace and I couldn't expect to see the genome of our child, to examine in fine detail how her DNA combined into a new person. We might as well have imagined buying a nuclear submarine. Instead, our genetics counselor performed a kind of verbal genome sequencing. She asked us about our families. The stories we told her gave her hints about whether mutations lurked in our chromosomes that might combine into dangerous possibilities in our child.

Grace's story was quick: Irish, through and through. Her ancestors had arrived in the United States in the early twentieth century, from Galway on one side, Kerry and Derry on the other. My story, as far as I understood it, was a muddle. My father was Jewish, and his family had come from eastern Europe in the late 1800s. Since Zimmer was German, I assumed he must have some German ancestry, too. My mother's family was mostly English with some German mixed in, and possibly some Irish—although a bizarre family story clattered down through the generations that we actually had a Welsh ancestor who only claimed to be Irish, for reasons lost to history. Oh, I added, someone on my mother's side of the family had definitely come over on the *Mayflower*. I was under the impression that he fell off the ship and had to get fished out of the Atlantic.

As I spoke, I could feel my smugness dissolving at its margins. What did I really know about the people who had come before me? I could barely remember their names. How could I know anything about what I had inherited from them?

Our counselor explained that my Jewish ancestry might raise the possibility of Tay-Sachs disease, a nerve-destroying disorder caused by inheriting two mutant copies of a gene called HEXA. The fact that my mother wasn't Jewish lowered the odds that I had the mutation. And even if I did, Grace's Irish ancestry probably meant we had nothing to worry about.

The more we talked about our genes, the more alien they felt to me. My mutations seemed to flicker in my DNA like red warning lights. Maybe one of the lights was on a copy of my HEXA gene. Maybe I had others in genes that scientists had yet to name, but could still wreak havoc on our child. I had willingly become a conduit for heredity, allowing the biological past to make its way into the future. And yet I had no idea of what I was passing on.

Our counselor kept trying to flush out clues. Did any relatives die of cancer? What kind? How old were they? Anyone have a stroke? I tried to build a medical pedigree for her, but all I could recall were secondhand stories. I recalled William Zimmer, my father's father, who died in his forties from a heart attack—I think a heart attack? But didn't an old cousin once tell me about rumors of overwork and despair? His wife, my grandmother, died of some kind of cancer, I knew. Was it her ovaries, or her lymph nodes? She had died years before I was born, and no one had wanted to burden me as a child with the oncological particulars.

How, I wondered, could someone like me, with so little grasp of his own heredity, be permitted to have a child? It was then, in a panic, that I recalled an uncle I had never met. I didn't even know he existed, until I was a teenager. One day my mother told me about her brother, Harry, how she would visit Harry's crib every morning to say hello. One morning, the crib was empty.

The story left me flummoxed, outraged. It wouldn't be until I was much older that I'd appreciate how doctors in the 1950s ordered parents to put children like Harry in a home and move on with their lives. I had no grasp of the awkward shame that would make those children all the more invisible.

I tried to describe Uncle Harry to our genetics counselor, but I might as well have tried sketching a ghost. As I blathered on, I convinced myself that our child was in danger. Whatever Harry had inherited from our ancestors had traveled silently into me. And from me it had traveled to my child, in whom it would cause some sort of disaster.

The counselor didn't look worried as I spoke. That irritated me. She asked me if I knew anything about Harry's condition. Was it fragile X? What did his hands and feet look like?

I had no answers. I had never met him. I had never even tried to track him down. I suppose I had been frightened of him gazing at me as he would at any stranger. We might share some DNA, but did we share anything that really mattered?

"Well," the counselor said calmly, "Fragile X is carried on the X chromosome. So we don't have to worry about that."

Her calmness now looked to me like sheer incompetence. "How can you be so sure?" I asked.

"We would know," she assured me.

"How would we know?" I demanded.

The counselor smiled with the steadiness of a diplomat meeting a dictator. "You'd be severely retarded," she said.

She started to draw again, just to make sure I understood what she was saying. Women have two X chromosomes, she explained, and men have one X and one Y. A woman with a fragile X mutation on one copy of her X chromosome will be healthy, because her other X chromosome can compensate. Men have no backup. If I carried the mutation, it would have been obvious from when I was a baby.

I listened to the rest of her lesson without interrupting.

A few months later, Grace gave birth to our child, a girl as it turned out. We named her Charlotte. When I carried her out of the hospital in a baby seat, I couldn't believe that we were being entrusted with this life. She didn't display any sign of a hereditary disease. She grew and thrived. I looked for heredity's prints on Charlotte's clay. I inspected her face, aligning photos of her with snapshots of Grace as a baby. Sometimes I thought I could hear heredity. To my ear, at least, she has her mother's laugh.

As I write this, Charlotte is now fifteen. She has a thirteen-year-old sister named Veronica. Watching them grow up, I have pondered heredity even more. I wondered about the source of their different shades of skin color, the tint of their irises, Charlotte's obsession with the dark matter of the universe, or Veronica's gift for singing. ("She didn't get that from me." "Well, she certainly didn't get it from *me.*")

Those thoughts led me to wonder about heredity itself. It is a word that we all know. Nobody needs an introduction to it, the way we might to *meiosis* or *allele.* We all feel like we're on a first-name basis with heredity. We use it to make sense of some of the most important parts of our lives. Yet it means many different things to us, which often don't line up with each other. Heredity is why we're like our ancestors. Heredity is the inheritance of a gift, or of a curse. Heredity defines us through our biological past. It also gives us a chance at immortality by extending heredity into the future.

I began to dig into heredity's history, and ended up in an

underground palace. For countless millennia, humans have told stories about how the past gave rise to the present, how people resemble their parents—or, for some reason, do not. And yet no one used the word *heredity* as we do today before the 1700s. The modern concept of heredity, as a matter worthy of scientific investigation, didn't gel for another century after that. Charles Darwin helped turn it into a scientific question, a question he did his best to answer. He failed spectacularly. In the early 1900s, the birth of genetics seemed to offer an answer at last. Gradually, people translated their old notions and values about heredity into a language of genes. As the technology for studying genes grew cheaper and faster, people became comfortable with examining their own DNA. As genetics passed its century mark, people began to order genetic tests to link themselves to missing parents, to distant ancestors, to racial identities. Genes became the blessing and the curse that our ancestors bestowed on us.

But very often genes cannot give us what we really want from heredity. Each of us carries an amalgam of fragments of DNA, stitched together from some of our many ancestors. Each piece has its own ancestry, traveling a different path back through human history. A particular fragment may sometimes be cause for worry, but most of our DNA influences who were are—our appearance, our height, our penchants—in inconceivably subtle ways.

While we may expect too much from our inherited genes, we also don't give heredity the full credit it's due. We've come to define heredity purely as the genes that parents pass down to their children. But heredity continues within us, as a single cell gives rise to a pedigree of trillions of cells that make up our entire bodies. And if we want to say we inherit genes from our ancestors—using a word that once referred to kingdoms and estates—then we should consider the possibility that we inherit other things that matter greatly to our existence, from the microbes that swarm our bodies to the technology we use to make life more comfortable for ourselves. We should try to redefine the word *heredity*, to create a more generous definition that's closer to nature than to our demands and fears.

I woke up one bright September morning and hoisted Charlotte, now two months old, from her crib. As Grace caught up on her sleep, I carried Charlotte to the living room, trying to keep her quiet. She was irascible, and the only way I could calm her was to bounce her in my arms. To fill the morning hours, I kept the television on: the chatter of local news and celebrity trivia, the pleasant weather forecast, a passing report of a small fire in an office at the World Trade Center.

Having been a father for all of two months had made me keenly aware of the ocean of words that surrounded my family. They flowed from our television and from the mouths of friends, they looked up from newspapers and leaped down from billboards. For now, Charlotte could not make sense of these words, but they were washing over her anyway, molding her developing brain to take on the capacity for language. She would inherit English from us, along with the genes in her cells.

She would inherit a world as well, a human-shaped environment that would help determine the opportunities and limits of her life. Before that morning, I felt familiar with that world. It would boast brain surgery and probes headed for Saturn. It would also be a world of increasing pavement and decreasing forests. But the fire grew that morning, and the television hosts mentioned reports that a plane had crashed into it. I rocked Charlotte as the television wove between ads and cooking tips and a second plane crashing into the second tower. The day mushroomed into catastrophe.

Charlotte's fussing had faded into sleepy comfort. She looked up at me and I down at her. I realized how consumed I had become with wondering what versions of DNA she might have inherited from me. I kept my arms folded tightly around her, wondering now what sort of world she was inheriting.

Notes:

"most wondrous map": National Human Genome Research Institute 2000.
genetic material pieced together from a mix of people: Wade 2002.
a gene called HEXA: US National Library of Medicine 2017